Sunantha Sodsee · Phayung Meesad ·
Watchareewan Jitsakul · Sakchai Tangwannawit
Editors

Proceedings of the 22nd International Conference on Computing and Information Technology (IC2IT2026)

Springer

Editors
Sunantha Sodsee
Faculty of Information Technology
and Digital Innovation
King Mongkut's University of Technology
North Bangkok
Bangkok, Thailand

Phayung Meesad
Faculty of Information Technology
and Digital Innovation
King Mongkut's University of Technology
North Bangkok
Bangkok, Thailand

Watchareewan Jitsakul
Faculty of Information Technology
and Digital Innovation
King Mongkut's University of Technology
North Bangkok
Bangkok, Thailand

Sakchai Tangwannawit
Faculty of Information Technology
and Digital Innovation
King Mongkut's University of Technology
North Bangkok
Bangkok, Thailand

ISSN 2367-3370 ISSN 2367-3389 (electronic)
Lecture Notes in Networks and Systems
ISBN 978-3-032-28047-3 ISBN 978-3-032-28048-0 (eBook)
https://doi.org/10.1007/978-3-032-28048-0

This Springer imprint is published by the registered company Springer Nature Switzerland AG
The registered company address is: Gewerbestrasse 11, 6330 Cham, Switzerland

Lecture Notes in Networks and Systems 5590

The series "Lecture Notes in Networks and Systems" publishes the latest developments in Networks and Systems—quickly, informally and with high quality. Original research reported in proceedings and post-proceedings represents the core of LNNS.

Volumes published in LNNS embrace all aspects and subfields of, as well as new challenges in, Networks and Systems.

The series contains proceedings and edited volumes in systems and networks, spanning the areas of Cyber-Physical Systems, Autonomous Systems, Sensor Networks, Control Systems, Energy Systems, Automotive Systems, Biological Systems, Vehicular Networking and Connected Vehicles, Aerospace Systems, Automation, Manufacturing, Smart Grids, Nonlinear Systems, Power Systems, Robotics, Social Systems, Economic Systems and other. Of particular value to both the contributors and the readership are the short publication timeframe and the world-wide distribution and exposure which enable both a wide and rapid dissemination of research output.

The series covers the theory, applications, and perspectives on the state of the art and future developments relevant to systems and networks, decision making, control, complex processes and related areas, as embedded in the fields of interdisciplinary and applied sciences, engineering, computer science, physics, economics, social, and life sciences, as well as the paradigms and methodologies behind them.

Indexed by SCOPUS, EI Compendex, INSPEC, WTI Frankfurt eG, zbMATH, SCImago.

All books published in the series are submitted for consideration in Web of Science.

For proposals from Asia please contact Aninda Bose (aninda.bose@springer.com).

Preface

In the present, the advancement of technology takes a significant role in motivating the economy, community, industry, and even daily life. The demand for high-skilled digital literacy workforces is increasing, which will drive the innovation industry to enhance economic value. This shows the growing importance of research and knowledge exchange in the digital technology field.

This year marks the 22nd International Conference on Computing and Information Technology (IC2IT), held on May 14–15, 2026, in Bangkok, Thailand. The conference brings together researchers, academicians, and experts from around the world to share their experiences, knowledge, and findings. It also provides opportunities to build collaborative networks and develop practical applications of cutting-edge research.

Herein, all papers in this book were reviewed by at least three qualified reviewers with expertise in artificial intelligence, machine learning, natural language processing, image processing, robotics, data science, cybersecurity, and innovative digital technologies.

Last but not least, the IC2IT committee hopes this book shows its value to researchers, academicians, experts, and those who pursue improvement, to enhance innovation nationally and internationally. The committee would like to thank all researchers, experts, reviewers, keynote speakers, and supporting organizations. Special thanks go to the Faculty of Information Technology and Digital Innovation at King Mongkut's University of Technology North Bangkok for their hard work in making this conference a success.

February 2026

Phayung Meesad
Sunantha Sodsee
Watchareewan Jitsakul
Sakchai Tangwannawit

Organization

Program Committee

S. Auwatanamongkol	NIDA, Thailand
T. Bernard	Li-Parad, France
S. Boonkrong	SUT, Thailand
N. Chumuang	MCRU, Thailand
T. Eggendorfer	TH Ingolstadt, Germany
S. Gammanee	KRU, Thailand
S. Hengpraprohm	NPRU, Thailand
S. Hiranpongsin	UBU, Thailand
W. Janratchakool	RMUTT, Thailand
T. Jensuttiwetchakul	KMUTNB, Thailand
W. Jitsakul	KMUTNB, Thailand
T. Kaewkiriya	BU, Thailand
S. Krootjohn	KMUTNB, Thailand
P. Kropf	Uni NE, Switzerland
M. Kubek	GSU, USA
K. Laosen	PSU, Thailand
U. Lechner	UniBW, Germany
N. Mata	BRU, Thailand
P. Meesad	KMUTNB, Thailand
S. Nuanmeesri	SSRU, Thailand
W. Pongpech	NIDA, Thailand
L. Poomhiran	SSRU, Thailand
K. Saengkrajang	PCRU, Thailand
P. Saengsiri	TISTR, Thailand
W. Saetang	PSU, Thailand
T. Sarawong	RMUTK, Thailand
S. Satiman	UBRU, Thailand
S. Sodsee	KMUTNB, Thailand
G. Somprasertsri	MSU, Thailand
W. Sriurai	UBU, Thailand
T. Sucontphunt	NIDA, Thailand
P. Tangwannawit	PCRU, Thailand
S. Tangwannawit	KMUTNB, Thailand
J. Tengsuthiwat	KMUTNB, Thailand
C. Thaenchaikun	PSU, Thailand

N. Utakrit	KMUTNB, Thailand
N. Utakrit	KMUTNB, Thailand
K. Viriyapant	KMUTNB, Thailand
P. Wongchampa	KMUTNB, Thailand
B. Yodwong	KMUTNB, Thailand

Additional Reviewers

Organizing Partners

In Cooperation with
Council of IT Deans of Thailand (CITT)
King Mongkut's University of Technology North Bangkok (KMUTNB)
FernUniversität in Hagen, Germany (FernUni)
Oklahoma State University, USA (OSU)
Edith Cowan University, Western Australia (ECU)
Hanoi National University of Education, Vietnam (HNUE)
Mahasarakham University (MSU)
National Institute of Development Administration (NIDA)
Rajamangala University of Technology Thanyaburi (RMUTT)
Kanchanaburi Rajabhat University (KRU)
Ubon Ratchathani University (UBU)

Contents

Bridging Data–Policy Gap in Universal Health Coverage: A Governance-Embedded Reference Architecture for Data-Driven Health Systems

Khachon Mongkonchoo(✉), Nalinpat Bhumpenpein, and Nathaporn Utakrit

King Mongkut's University of Technology North Bangkok, Bangkok, Thailand
s6607011956020@email.kmutnb.ac.th, nalinpat.b@itd.kmutnb.ac.th

Abstract. As countries pursue sustainable Universal Health Coverage (UHC), the scale and heterogeneity of health data have increased substantially. However, a persistent data–policy gap remains, in which decision-makers are unable to operationalize available data for strategic planning and resource allocation due to architectural fragmentation and weak governance enforcement. This study conducts a Systematic Literature Review (SLR) of 45 empirical studies selected from 1,686 records published between 2010 and 2024 and synthesizes the findings as design evidence for constructing a governance-embedded reference architecture for data-driven UHC systems. Based on cross-country implementation evidence, we propose a five-layer federated reference architecture comprising: (i) a data source layer for system integration; (ii) an interoperability and middleware layer utilizing FHIR-based APIs; (iii) a governance and trust layer embedding policy rules via smart contracts; (iv) an analytics and intelligence layer for ML-driven decision support; and (v) a policy and application layer for strategic purchasing and monitoring. The paper contributes to a set of architecture design principles and an evaluation framework for assessing data–policy alignment, scalability, and governance compliance in large-scale health information systems. Using the PPTDG framework (People, Process, Technology, Data, Governance) as an analytical lens, the findings indicate that entrenched data silos and governance deficits represent primary system-level constraints, often outweighing purely technical limitations. The analysis highlights a strategic shift toward federated architectures with open API–based interoperability and Platform-as-Governance mechanisms, such as blockchain-enabled audit trails, to enable automated policy enforcement and institutional trust. The study also emphasizes the role of Data Translators as socio-technical interfaces that align analytical outputs with policy execution and proposes a conceptual foundation for a Strategic Analytic Framework integrating financial incentives with health informatics.

Keywords: Universal Health Coverage · Data-Policy Gap · Federated Architecture · Data Translators · Strategic Analytic Framework

S. Sodsee et al. (Eds.): IC2IT 2026, LNNS 5590, pp. 1–10, 2026.
https://doi.org/10.1007/978-3-032-28048-0_1

1 Introduction

The global digital transformation of healthcare systems has introduced large-scale information infrastructures intended to support the management and sustainability of Universal Health Coverage (UHC). Contemporary health information ecosystems including electronic medical records (EMRs), national medical claims platforms, and population health databases generate high-volume, high-velocity, and heterogeneous data streams. In principle, these data assets enable data-driven policy functions such as cost containment, strategic purchasing, quality monitoring, and pandemic response. In practice, however, many low- and middle-income countries (LMICs), as well as several high-income health systems, exhibit a persistent "data-rich but information-poor" condition. Despite continuous data generation in operational systems, policy decisions often rely on fragmented and outdated information due to limited integration across systems. This disconnect creates inefficiency, as health security purchasers lack timely and comprehensive data to enable strategic purchasing, resulting in increased risk of waste, misuse, and fraud. During major public health crises like COVID-19, the absence of real-time data sharing across institutions limits the ability to quickly adapt policies amid uncertainty.

Most research addresses technical interoperability and governance frameworks separately. While individual studies examine data standards, health information exchange, or policy tools, integrated models embedding policy logic into health platforms are rare. This is due to a mismatch: data systems handle transactions, but policy needs ongoing aggregation, enforcement, and feedback. Consequently, current systems show weak links between data processes and governance logic. This paper reframes the data–policy divide as an information system architecture issue and proposes a governance-integrated reference architecture for UHC platforms. Using the PPTDG framework (People, Process, Technology, Data, Governance), it explores key barriers to effective data use in policy execution. (2) What architectural approaches can mitigate fragmentation and enable policy-aware interoperability? and (3) What technical and organizational enablers are required to sustain such architectures at scale?

2 Methodology

This study adopts a Design Science Research (DSR) methodology to develop a governance-embedded reference architecture for data-driven UHC systems. The primary contribution is an architectural artifact designed to address a well-defined socio-technical problem in large-scale health information systems. Structured around three core activities—problem identification, design and development, and evaluation—the research frames the persistent data–policy gap in UHC as an architectural challenge characterized by fragmented infrastructures and weak governance enforcement. Rather than proposing isolated technical fixes, the study aims to synthesize empirical evidence into a coherent system-level architecture capable of embedding policy logic directly into information platforms.

2.1 Knowledge Base and Requirements Elicitation

The DSR knowledge base was constructed through a Systematic Literature Review (SLR) following PRISMA 2020 guidelines to ensure transparency. Literature was

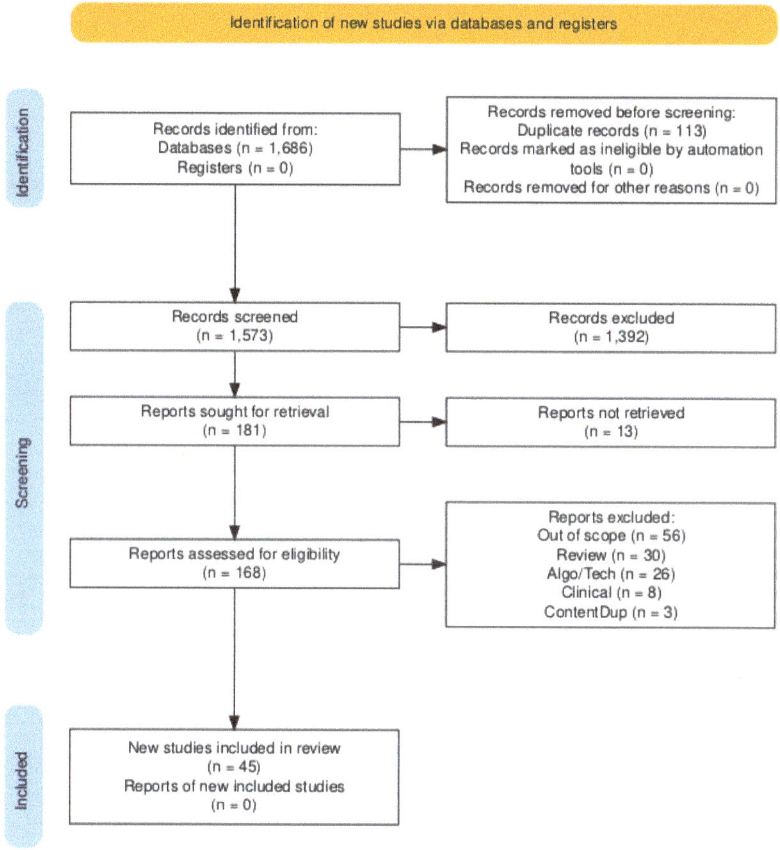

Fig. 1. PRISMA 2020 flow diagram illustrating the literature selection process.

retrieved from 5 major academic databases—PubMed, IEEE Xplore, ACM Digital Library, ScienceDirect, and SpringerLink—using Boolean queries related to UHC, health policy, and interoperability platforms. Out of 1,686 initial records, 45 full-text studies were chosen for meeting strict criteria on empirical system-level deployment. To ensure a system-level focus, standalone clinical applications and studies lacking implementation context were excluded. The selection process is illustrated in the PRISMA flow diagram (Fig. 1), which was generated using the PRISMA2020 tool [1].

2.2 Design and Development of the Artifact

The extracted findings are synthesized into a governance-embedded federated reference architecture, constituting the primary DSR artifact. Architectural design decisions are derived by mapping PPTDG-identified constraints to corresponding system components and control mechanisms. The resulting architecture decomposes the UHC information ecosystem into five logical layers: data sources, interoperability middleware, governance and trust, analytics, and policy applications. This reference model defines logical interaction patterns that can be instantiated across heterogeneous national health system contexts, balancing architectural generalizability with practical relevance.

3 Finding

The architectural artifact produced through the DSR process is a governance-embedded, multi-layer federated reference architecture for data-driven UHC systems, explicitly designed to address the systemic data–policy gap by integrating interoperability, governance enforcement, and analytics-driven policy execution within a unified information system model. Instead of replacing old systems, the architecture outlines system roles and interactions that can be used in various national health contexts (Fig. 2).

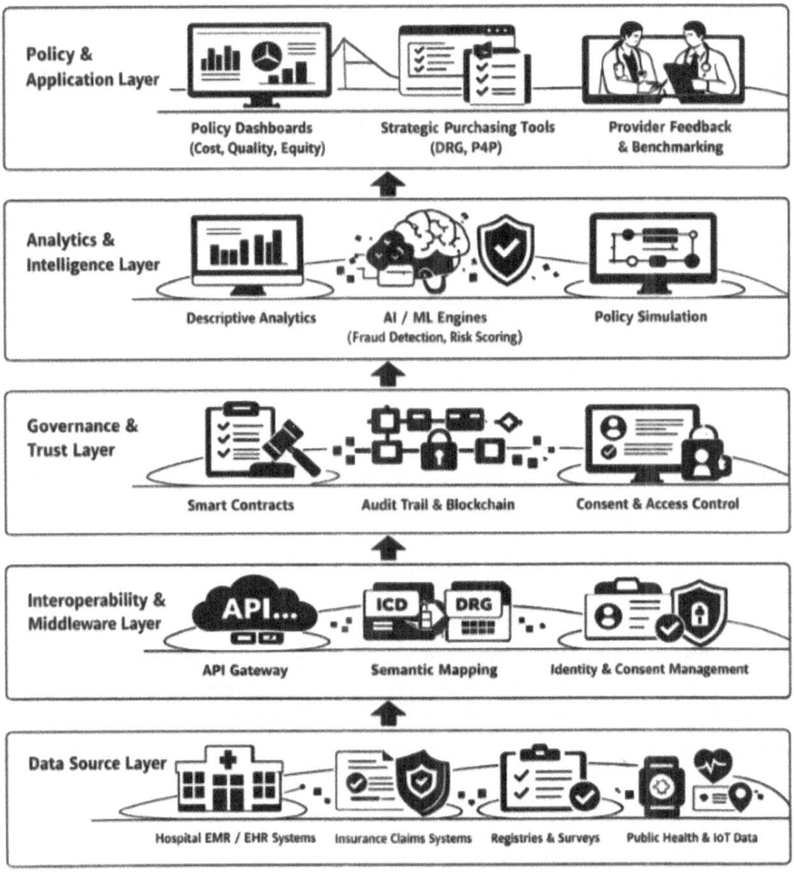

Fig. 2. Governance-Embedded Health Data Architecture.

The architecture is organized into five logically decoupled yet functionally integrated layers, enabling traceability between empirically observed constraints—identified through the SLR and structured using the PPTDG framework. The Data Source

Layer preserves data sovereignty across heterogeneous systems, insurance claims platforms, registries, surveys, and public health or IoT sources, while the Interoperability and Middleware Layer enables virtual integration via API gateways, FHIR-based interfaces, semantic mapping, and identity and consent management to achieve syntactic and semantic interoperability with runtime privacy enforcement. The Governance and Trust Layer embed policy logic as executable system components—such as rule engines, smart contracts, and blockchain-enabled audit trails—transforming governance from post-hoc oversight into a runtime capability that enforces reimbursement, eligibility, and accountability requirements. This layer complies with data protection regulations like PDPA and GDPR using a Privacy-by-Design approach. By decoupling identity management and implementing blockchain-based audit trails, it guarantees transparency and patient consent in data handling, reducing risks related to sensitive health information. Above this, the Analytics and Intelligence Layer operationalizes federated data through descriptive analytics, AI/ML-based fraud detection and risk scoring, and policy simulation, with outputs constrained by governance rules to ensure policy-aware decision support. Finally, the Policy and Application Layer delivers these capabilities through dashboards, strategic purchasing tools, and provider feedback and benchmarking interfaces, thereby closing the data–policy feedback loop and enabling accountable, incentive-aligned, and evidence-based UHC management.

3.1 Systemic Barriers to Data-Driven Policy Execution

Evidence from a PRISMA-compliant review shows that structural and governance flaws, rather than technical issues, are the main barriers to data-driven Universal Health Coverage (UHC). Structural fragmentation frequently manifests as isolated "information islands"; for instance, China's historical medical insurance evolution produced over 2,000 fragmented county-level silos that significantly hindered the operationalization of national reimbursement policies [2]. Similarly, Germany's "analog mandate" and segmented healthcare landscape created substantial resistance to electronic record adoption, as digital processes struggled to mirror complex, legacy bureaucratic paper workflows [3]. In the United Kingdom, persistent fragmentation across primary, secondary, and social care continues to result in critical information gaps during patient transitions, thereby impeding integrated care initiatives [4]. This lack of architectural harmony is further evidenced in Russia's Far East, where incompatible EMR implementations complicated inter-hospital cooperation and underscored the vital necessity for unified standards [5]. Beyond structural silos, governance deficits cultivate environments of mistrust and data hoarding; private providers in Ukraine often withhold data from national systems due to missing incentives and regulatory apprehension, creating significant planning "blind spots" [6]. In decentralized models such as Norway's, the "One Citizen–One Journal" initiative faced protracted delays caused by tensions between central authority and municipal autonomy [7]. Furthermore, the absence of unified data privacy and consent frameworks, as observed in India, has historically hampered national health infrastructure development [8], while the UAE's pre-pandemic digitalization gap constrained effective crisis management [9]. Finally, a persistent "data–policy disconnect" exists even where high-quality data is available; in Thailand, the misalignment between National Health Examination Survey (NHES) cycles and policy requirements led decision-makers to favor more accessible, yet less accurate, routine facility datasets [10]. This is echoed in Ethiopia, where limited local data utilization in immunization programs stems from a lack of accountability and feedback mechanisms at the district level [11].

3.2 Architectural Strategies for Policy-Aware Interoperability

To mitigate these systemic barriers, literature advocates for a transition from monolithic infrastructures toward flexible, federated, and intelligent architectures. India's "Health Stack" exemplifies this shift by decoupling the consent layer from data flow [8], while the Netherlands' ELAN infrastructure utilizes virtual integration and pseudonymization to enable population health management without consolidating sensitive clinical data [12]. Interoperability requires semantic standards like ICD-10, SNOMED-CT, and FHIR. West China Hospital unified data from 134 different systems using strict master data governance [13]. Similarly, the PeOPLe platform in Japan leverages ICT and unique identifiers to unify life-course data [14], supported by frameworks emphasizing standardized impact assessments [15]. Conversely, the lack of standard terminology remains a primary barrier in the Russian Far East [5]. To resolve trust deficits, blockchain technology offers immutable ledgers for claims adjudication, reducing fraud in the US Medicare system [16] and the broader insurance industry [17]. Decentralized marketplaces for patient-generated data empower individuals to monetize and control their information while ensuring provenance [18], and security frameworks using digital identities like MOSIP ensure privacy-preserving exchange across heterogeneous systems [19]. Advanced platforms increasingly integrate Artificial Intelligence (AI) for strategic purchasing, as demonstrated by Taiwan's NHIA, which saved billions by detecting duplicate claims and abnormal patterns [20]. AI-driven triage in China's internet hospitals [21], pandemic preparedness tools in Africa [22], big-data models for hearing health policy [23], and IRA-NRM analysis for resource allocation [24] highlight the technology's diverse utility, including cloud-based data mining for maternal care in the Philippines [25]. Furthermore, architectures must support evolving reimbursement models, such as Diagnosis-Related Groups (DRG) in China [26] and Pay-for-Performance (P4P) in Taiwan [27]. Switzerland's needs-based hospital planning [28] and US Medicaid's telehealth evaluations [29] illustrate the role of granular data in assessing clinical appropriateness. Finally, platforms must exhibit resilience through emergency collaborative technologies [30], disaster preparedness as seen in Japan [31], and telemedicine for smart aging in Hong Kong [32].

3.3 Technical and Organizational Enablers for Scalable Architectures

The sustainability of health information architectures depends on critical socio-technical enablers that align people, processes, and incentives. Leadership serves as a primary determinant; in Tanzania, the IMIS implementation succeeded only where district medical officers actively championed the system [33], a finding echoed in the "innovation-system fit" required for Asian critical care registries [34]. Furthermore, "Data Translators" are essential to bridge the socio-technical gap, as high-quality data fails to influence policy without intermediaries to translate technical outputs into strategic insights [10], necessitating the alignment of digital strategies with global frameworks and local leadership [35]. Financial sustainability must be ensured by linking data quality to incentives; Taiwan's P4P programs [27] and medical device regulations [36] demonstrate how policy directly impacts market availability. However, incentive design must be carefully calibrated to avoid compromising service quality, as observed in China's DRG implementation [26]. Granular data further supports Switzerland's "needs-based" planning [28] and US Medicaid reimbursement decisions [29], whereas the withdrawal of UHC coverage in South Korea illustrated the direct impact of financial policies on fatality rates [37]. Switzerland's psychiatric day care tariffs further confirm the necessity of accurate data for financial calculation [38]. Moreover, systems must deliver intrinsic value to producers through user-centric designs that simplify workflows, as seen in Tanzania [33]. A feedback-driven approach is vital to avoid the accountability gaps identified in Ethiopia [11]. Performance measurement frameworks for community health workers must balance data demand with reporting burdens [39], ensuring alignment with clinical workflows to drive adoption in Quebec [40], and utilizing policy supports for implementation across Europe [41]. Ultimately, robust monitoring and evaluation (M&E) mechanisms—such as M&E production frontiers [42], WHO inequality monitoring [43], and real-time data for decision-making under uncertainty [44]—are indispensable. Economic evaluations, including cost-utility analyses of Singapore's NGEMR [45] and drug utilization monitoring in China [46], provide the evidence-based justification for the long-term value and equity of integrated systems.

4 Discussion

The proposed architecture adopts a federated, multi-layer design in which data remains at its source while governance and analytics are enforced through interoperable middleware, enabling real-time alignment between data generation and policy decision-making. This synthesis suggests a paradigm shift toward "Platform as Governance," where the analytic platform transcends being a mere tool to becoming the governance mechanism itself. By embedding rules, standards, and reimbursement logic directly into the system logic—such as via Smart Contracts in blockchain [16, 18] or validation rules in API gateways [8]—the platform automatically enforces transparency and accountability, making the "rules of the game" explicit and immutable. To bridge the data-policy gap without overhauling legacy systems, a "Middle-ware" approach is essential. This involves building an interoperability layer that connects disparate systems, as illustrated by the regional infrastructures in the UK [4] and Singapore's integrated clusters [45], thereby reducing duplicate testing and systemic fragmentation. For UHC systems, the platform must support the "purchaser-provider split" by offering granular visibility into cost and quality—exemplified by Switzerland's capacity planning [28] and Taiwan's AI-based claim reviews [20]—while maintaining provider autonomy. The integration of financial and clinical data remains the keystone for enabling value-based healthcare. Ultimately, the federated nature of this model enhances implementation feasibility through incremental, virtual integration. By prioritizing Governance (G) and People (P) within the PPTDG framework, it offers a sustainable and scalable pathway for resource-constrained systems to bridge the data–policy gap effectively without a centralized warehouse overhaul.

5 Conclusion

Bridging the data-policy gap in UHC requires a holistic strategy that transcends technology. This study conceptualizes the data–policy gap as a system architecture problem and proposes a governance-embedded reference architecture for UHC platforms. While technical solutions like Federated APIs, AI, and Blockchain offer powerful capabilities, their success depends on Governance, Incentives and People (data translators, leadership).

While this study proposes a comprehensive Governance-Embedded Reference Architecture, future research should focus on the empirical validation of this framework within a specific national health security context. Subsequent work will involve developing a functional prototype of the governance enforcement layer to evaluate its effectiveness in automating policy compliance and reducing administrative delays in UHC resource allocation.

References

1. Haddaway, N.R., et al.: PRISMA2020: an R package and Shiny app for producing PRISMA 2020-compliant flow diagrams. Campbell Syst. Rev. **18**, e1230 (2022). https://doi.org/10.1002/cl2.1230
2. Li, Y., Lu, C., Liu, Y.: Medical insurance information systems in China: mixed methods study. JMIR Med. Inform. **8**, e18780 (2020). https://doi.org/10.2196/18780
3. Pohlmann, S., et al.: Digitalizing health services by implementing a personal electronic health record in Germany. J. Med. Internet Res. **22**, e15102 (2020). https://doi.org/10.2196/15102
4. Elizondo, A., et al.: Implementing integrated care infrastructure: a longitudinal study on NHS England. Health Policy **160**, 105237 (2025). https://doi.org/10.1016/j.healthpol.2024.105237
5. Jung, S.Y., et al.: Barriers and facilitators to implementation of nationwide electronic health records in the Russian Far East. Int. J. Med. Inf. **143**, 104244 (2020). https://doi.org/10.1016/j.ijmedinf.2020.104244
6. Hellowell, M., et al.: Private provision of health services in Ukraine: a qualitative exploration of current policy challenges. Health Policy **158**, 105349 (2025). https://doi.org/10.1016/j.healthpol.2025.105349
7. Linstad, L.H., et al.: Investigating the Norwegian eHealth governance model: document study. J. Med. Internet Res. **26**, e59717 (2024). https://doi.org/10.2196/59717
8. Balsari, S., et al.: Reimagining health data exchange: an application programming interface-enabled roadmap for India. J. Med. Internet Res. **20**, e10725 (2018). https://doi.org/10.2196/10725
9. Pillai, N., et al.: Impact of digitalization of the healthcare industry and covid 19 Management: case of the UAE. In: Proceedings of 12th ICCCNT, pp. 1–9. IEEE (2021). https://doi.org/10.1109/ICCCNT51525.2021.9580088
10. Tantivess, S., et al.: Utilisation of evidence from Thailand's national health examination survey in policy development. Health Res. Policy Syst. **17**, 104 (2019). https://doi.org/10.1186/s12961-019-0512-4
11. Tilahun, B., et al.: Using health data for decision-making at each level of the health system to achieve universal health coverage in Ethiopia. Health Res. Policy Syst. **19**, 48 (2021). https://doi.org/10.1186/s12961-021-00694-1
12. Ardesch, F.H., et al.: The introduction of a data-driven population health management approach in the Netherlands since 2019. Health Policy **132**, 104769 (2023). https://doi.org/10.1016/j.healthpol.2023.104769

13. Wang, M., et al.: Big data health care platform with multisource heterogeneous data integration and massive high-dimensional data governance. JMIR Med. Inf. **10**, e36481 (2022). https://doi.org/10.2196/36481
14. Nomura, S., et al.: Contributions of information and communications technology to future health systems and Universal Health Coverage. Health Res. Policy Syst. **18**, 73 (2020). https://doi.org/10.1186/s12961-020-00585-x
15. Chouvarda, I., et al.: Connected health services: framework for an impact assessment. J. Med. Internet Res. **21**, e14005 (2019). https://doi.org/10.2196/14005
16. Mackey, T.K., et al.: Combating health care fraud and abuse: conceptualization and prototyping study of a blockchain antifraud framework. J. Med. Internet Res. **22**, e18623 (2020). https://doi.org/10.2196/18623
17. Buksh, Z., et al.: Blockchain-powered fraud prevention in health insurance industry. In: Proceedings of 6th ACM B-SCI, pp. 1–8. ACM (2024). https://doi.org/10.1145/3659463.3660019
18. Subramanian, H.: A decentralized marketplace for patient-generated health data: design science approach. J. Med. Internet Res. **25**, e42743 (2023). https://doi.org/10.2196/42743
19. Shrivastava, S., Srikanth, T.K.: A framework for secure and privacy preserving health data exchange across health information systems. In: Proceedings of 15th ICEGOV, pp. 41–49. ACM (2022). https://doi.org/10.1145/3560107.3560115
20. Tai, H.Y.: Applications of big data and AI. In: Lee, P.C., et al. (eds.) Digital Health Care in Taiwan, pp. 207–217. Springer, Heidelberg (2022). https://doi.org/10.1007/978-3-031-05160-9_11
21. Zhang, G.W., et al.: In-depth examination of the functionality and performance of the internet hospital information platform. J. Med. Internet Res. **26**, e54018 (2024). https://doi.org/10.2196/54018
22. Ibeneme, S., et al.: Data revolution, health status transformation and the role of AI for health and pandemic preparedness. BMC Proc. **15**, 22 (2021). https://doi.org/10.1186/s12919-021-00228-1
23. Katrakazas, P., et al.: A big-data informed model approach to hearing health policy decision making. In: Proceedings of 2018 SITIS, pp. 725–729. IEEE (2018). https://doi.org/10.1109/SITIS.2018.00116
24. Chen, P.T.: Medical big data applications: intertwined effects and effective resource allocation strategies. Technol. Forecast. Soc. Change **130**, 150–164 (2018). https://doi.org/10.1016/j.techfore.2018.01.033
25. Castillo, A.G., et al.: Cloud-based data mining framework: a model to improve maternal healthcare. In: Proceedings of 2nd ICCSP, pp. 21–28. ACM (2018). https://doi.org/10.1145/3199478.3199483
26. Wang, Y., et al.: The impacts of medical insurance payment methods on medical bills and service quality. J. Bus. Res. **169**, 114292 (2023). https://doi.org/10.1016/j.jbusres.2023.114292
27. Hsieh, H.M., et al.: Associations between participation in a diabetes P4P program and health outcomes in Taiwan. Gen. Hosp. Psychiatry **94**, 99–107 (2025). https://doi.org/10.1016/j.genhosppsych.2025.02.025
28. Bleibtreu, E., et al.: Service-, needs-, and quality-based hospital capacity planning in Switzerland. Health Policy **126**, 1277–1282 (2022). https://doi.org/10.1016/j.healthpol.2022.09.011
29. Saravanakumar, S., Ostrovsky, A.: Evaluation of telehealth services for reimbursement in the us medicaid population. J. Med. Internet Res. **26**, e46412 (2024). https://doi.org/10.2196/46412

30. Udo, I.J., Ekpenyong, M.E.: Improving emergency healthcare response using real-time collaborative technology. In: Proceedings of 4th ICMHI, pp. 165–173. ACM (2020). https://doi.org/10.1145/3418094.3418118
31. Moriwaki, Y., et al.: Analysis of pharmaceutical demand using national health insurance data. Healthcare **11**, 3029 (2023). https://doi.org/10.3390/healthcare11233029
32. Cheng, L., et al.: Towards data-driven tele-medicine intelligence: community-based mental healthcare paradigm shift. Health Inf. Sci. Syst. **11**, 14 (2023). https://doi.org/10.1007/s13755-022-00198-4
33. Schuetze, L., et al.: Factors affecting successful implementation of a digital intervention for health financing. J. Med. Internet Res. **25**, e38818 (2023). https://doi.org/10.2196/38818
34. Tolppa, T., et al.: Determinants of implementation of a critical care registry in Asia. J. Med. Internet Res. **25**, e41028 (2023). https://doi.org/10.2196/41028
35. Holl, F., et al.: Tanzania's and Germany's digital health strategies and WHO's global strategy. J. Med. Internet Res. **26**, e52150 (2024). https://doi.org/10.2196/52150
36. Peng, J.Y., et al.: Impact of reimbursement regulation on patient access to innovative medical devices in Taiwan. Value Health Reg. Issues **42**, 100978 (2024). https://doi.org/10.1016/j.vhri.2023.12.009
37. Moon, D., et al.: Universal health coverage saves more lives among severely ill COVID-19 patients. Health Res. Policy Syst. **18**, 116 (2024). https://doi.org/10.1186/s12961-024-01212-9
38. Noll, S., et al.: A new case-mix based payment system for psychiatric day care in Switzerland. Health Policy **131**, 104797 (2023). https://doi.org/10.1016/j.healthpol.2023.104797
39. Agarwal, S., et al.: A conceptual framework for measuring community health workforce performance. Hum. Res. Health **17**, 86 (2019). https://doi.org/10.1186/s12960-019-0422-0
40. Motulsky, A., et al.: Evaluation of a national e-booking system for medical consultation in a universal system. Health Policy **131**, 104759 (2023). https://doi.org/10.1016/j.healthpol.2023.104759
41. Tan, S., et al.: How do policy supports enable implementation of integrated care programs? Health Policy **160**, 105334 (2025). https://doi.org/10.1016/j.healthpol.2025.105334
42. Mukherjee, M., Batta, A.: Optimum monitoring and evaluation production frontier for risk-informed governance. J. Health Popul. Nutr. **43**, 148 (2024). https://doi.org/10.1186/s41043-024-00632-1
43. Hosseinpoor, A.R., Bergen, N.: Health inequality monitoring. In: Population Health Monitoring, pp. 151–173. Springer, Heidelberg (2019). https://doi.org/10.1007/978-3-319-76562-4_8
44. Hadjisotiriou, S., et al.: Decision making under deep uncertainty for pandemic policy planning. Health Policy **131**, 104831 (2023). https://doi.org/10.1016/j.healthpol.2023.104831
45. Chen, C., et al.: Economic evaluation of next generation EMR in Singapore: cost-utility analysis. J. Med. Internet Res. **27**, e70484 (2025). https://doi.org/10.2196/70484
46. Liu, Y., et al.: National Health Insurance Coverage policy and lenvatinib use for advanced HCC. Int. J. Equity Health **22**, 256 (2023). https://doi.org/10.1186/s12939-023-02052-9

TWIN-Net: An Innovative Multi-Feature Model for Assessing Levels of Students' Satisfaction in Asynchronous E-Learning Forums

Sulis Sandiwarno[1,2], Dana Indra Sensuse[1(✉)], Harry Budi Santoso[1], DedenSumirat Hidayat[3], Ally S. Nyamawe[4], Abdallah Yousif[5], Jahns Michael[1], and Wawan Gunawan[2]

[1] Universitas Indonesia, Kota Depok, Indonesia
sulis.sandiwarno@mercubuana.ac.id, {dana,harrybs}@cs.ui.ac.id, jahns.michael51@ui.ac.id

[2] Universitas Mercu Buana, West Jakarta, Indonesia
wawan.gunawan@mercubuana.ac.id

[3] National Research and Innovation Agency, Jakarta, Indonesia
dede025@brin.go.id

[4] United Nations University Institute in Macau, Macau SAR, China
nyamawe@unu.edu

[5] Sudan University of Technology, Khartoum, Sudan

Abstract. Information technology significantly supports learning processes through e-learning platforms. An essential element of an e-learning environment is the asynchronous discussion forum, which facilitates the exchange of ideas and feedback between students and lecturers. Machine learning models have recently been applied to evaluate e-learning student satisfaction by analyzing lecturer feedback. However, most of the existing approaches focus primarily on modeling the strong relationship between sentiment and semantic similarity. Additionally, previous studies have not examined students' emotional expressions or identified topics that elicit similar responses from lecturers. To address these limitations, we propose a novel model, called Textual Wide Integration Network (TWIN-Net), that unifies sentiment, semantic similarity, emotion, and topic features through a weighted fusion mechanism to assess student satisfaction in online discussions. The proposed architecture comprises four primary neural components. To construct encoders for sentiment, semantic similarity, emotion, and topic sentences, it integrates Bidirectional Long Short-Term Memory (BiLSTM) and Convolutional Neural Network (CNN) algorithms, leveraging the RoBERTa and BERTopic frameworks as foundational models. The experimental results indicate that TWIN-Net demonstrates a notable improvement over existing approaches, as evidenced by its average F1-score of 92.72%.

Keywords: Student Satisfaction · E-Learning · Deep Neural Network · Sentiment · Semantic Similarity · Emotion · Topic Modeling

S. Sodsee et al. (Eds.): IC2IT 2026, LNNS 5590, pp. 11–25, 2026.
https://doi.org/10.1007/978-3-032-28048-0_2

1 Introduction

The advances in information and network technology have reshaped how people access and share knowledge, with the internet becoming the primary platform for knowledge exchange. Digital services now span banking, governance, commerce, and education, where several higher education institutions have adopted e-learning to support skill development [1]. E-learning offers an alternative to traditional teaching methods by leveraging the internet and web technologies. Notably, Learning Management Systems (LMS), serve as digital platforms that facilitate instructional delivery and engagement [2]. These systems include collaborative features, such as asynchronous discussion forums, which support interaction and knowledge exchange between students and lecturers. To assess the effectiveness of e-learning systems, particularly online discussion forums, previous studies have employed machine learning models based on sentiment [3], emotion [4], and semantic similarity analysis [5]. Furthermore, earlier studies have employed topic analysis in online discussion forums to examine the depth and focus of participation in asynchronous discussions [6]. Although most of the approaches mentioned have been widely used, machine learning classifiers have typically adopted sentiment, emotion, semantic similarity, or topic analysis individually. Without integrating sentiment, emotion, semantic similarity, and topic analysis, it is difficult to fully capture the emotional nuances and thematic relationships in asynchronous interactions between lecturers and students. To address this gap, the present study introduces a novel framework, Textual Wide Integration Network(TWIN-Net), designed to evaluate student satisfaction within asynchronous online learning discussion forums based on lecturers' responses. The CNN [7], BiLSTM models [8] are further improved by the Robustly Optimized BERT Approach (RoBERTa), and BERTopic [9] to enhance the extraction of multiple features from student perspectives and lecturer feedback within the discussion. This study employs the emotional framework from Ekman [10], comprising happiness (positive) and anger, disgust, fear, and sadness (negative), to analyze users' emotions. The key contributions of the paper are as follows:

- Firstly, we propose TWIN-Net, a new model combining CNN, BiLSTM, and RoBERTa to extract key features from student-lecturer interactions. This represents a pioneering effort to jointly evaluate sentiment, emotions, semantic similarity, and topic analysis in asynchronous e-learning forums.
- Secondly, we provide a more in-depth analysis through a visualization that illustrates the interconnections among sentiment, emotion, semantic similarity, and topic analysis, aiming to better understand how students' viewpoints are reflected in lecturers' feedback in the asynchronous educational forums.

According to the contributions and motivations outlined previously, this research seeks to address the following research questions (RQs):

- RQ1: Does TWIN-Net demonstrate significant improvement over previous models for assessing students' satisfaction through the integration of sentiment, emotion, semantic similarity, and topic features?
- RQ2: Does the proposed TWIN-Net sustain stable performance across various embedding techniques, including GloVe, fastText, BERT, and RoBERTa?

The remainder of this paper is structured as follows. Section 2 reviews relevant existing methods that are based on machine learning techniques. Section 3 outlines the proposed methodology. Section 4 details the experimental design and presents the findings. Finally, Sect. 5 concludes the paper and discusses potential directions for future research.

2 Related Works

Recent studies show that text classification is essential in Natural Language Processing (NLP) for analyzing online learning interactions. Liu et al. [11] introduced BERT-CNN for identifying opinions in MOOC forums, while Koufakou [12] demonstrated that RoBERTa clearly outperforms traditional classifiers such as SVM and NB for sentiment and topic analysis. Several works further explored hybrid RoBERTa models: Paneru et al. [13] used RoBERTa-CNN, Semary et al. [14] proposed RoBERTa-CNN–LSTM, and Rahman et al. [15] combined RoBERTa with BiLSTM, all showing consistent performance gains over conventional deep learning. Lak et al. [16] integrated RoBERTa, BiLSTM, and self-attention to enhance contextual feature learning. In the education domain, Sandiwarno et al. [5] developed E-SATNet that combines CNN, BiLSTM, and BERT embeddings, which proved more effective than standard deep learning models for capturing sentiment and semantic relationships in online forum interactions.

Although previous research has shown promising results, a comprehensive framework that jointly incorporates sentiment, emotion, semantic similarity, and topic analysis, particularly within e-learning discussion forums, has yet to be developed. A lack of integrated analysis limits the ability to uncover emotional complexity and thematic cohesion within the dialogue between students and lecturers. As a result, current models only partially capture user satisfaction and interaction dynamics in asynchronous learning environments.

3 Proposed Method

This study presents TWIN-Net, a multi-branch representation learning framework designed to evaluate student satisfaction in asynchronous learning discussions, as shown in Fig. 1. The model combines four complementary information sources, including sentiment polarity, semantic similarity, emotional cues, and discussion topics extracted from student posts and lecturer responses. Each branch uses RoBERTa as the embedding backbone to capture contextualized representations, while different neural components extract specialized features relevant to each analytical dimension. The final output is produced through feature fusion and multi-class classification.

3.1 Input Embedding Layer

In the first stage of processing, TWIN-Net analyzes raw text from students and lecturers by mapping it to context-rich numerical representations. This is achieved through the RoBERTa language model, which excels at modeling semantic and syntactic relations beyond the capabilities of static embeddings such as Word2Vec or GloVe. The input embedding layer processes raw text from students and lecturers, transforming it into dense numerical representations. For instance, a student's post may be expressed as a series of tokens.

$$S = \{w_1, \ldots, w_n\} \tag{1}$$

and a lecturer's reply as

$$L = \{u_1, \ldots, u_m\} \tag{2}$$

Each token is converted into its subword representation using RoBERTa's Byte-Pair Encoding (BPE) tokenizer before entering the model. This tokenization ensures robust handling of domain-specific terms, informal spellings, and out-of-vocabulary words common in online discussion forums. After tokenization, both sequences are passed through RoBERTa to generate contextualized embeddings for each token. Unlike static embeddings, RoBERTa encodes token meaning by considering its full bidirectional context, enabling the model to interpret complex expressions such as "unclear explanation", "I'm confused", or "thank you, this helps" in varied situations. The contextualized embeddings for tokens S and L are represented as:

$$\Phi(S) = [h_1, h_2, h_3, \ldots.., h_n], \Phi(L) = \left[g_1, g_2, g_3, \ldots.., g_n\right] \tag{3}$$

where $h_i, g_j \in \mathbb{R}^d$ denote the bidirectional contextual representations for each token. These embeddings serve as the shared input for all downstream branches.

3.2 Sentiment Feature Extraction (CNN-RoBERTa)

To collect sentiment-related cues in the student's post, TWIN-Net builds on RoBERTa's contextual embeddings with a CNN model from previous approaches [17, 18]. This component's concept is that emotion is frequently represented by brief, localized verbal patterns, such as "very helpful", "confusing explanation", or "not clear", which can be efficiently recognized by convolutional filters. Given the contextual embedding sequence $\Phi(S)$, the CNN applies a sliding window of size k across consecutive token embeddings. Each window corresponds to an *n-gram* region within the sentence. The convolution operation for a given window is formalized in Eq. (4):

$$A = \underset{i}{max}\, \sigma\left(W_s * \Phi(S)_{i:i+k-1} + b_s\right) \tag{4}$$

where W_s represents the convolution filter, $\Phi(S)_{i:i+k-1}$ corresponds to the sequence of embeddings within the current window, representing a contiguous segment of the input text, b_s is the bias term, k is the kernel size, and σ is the non-linear activation (ReLU). After convolving through all feasible windows, *max-pooling* is used to choose the most powerful activation at each point. This pooling procedure ensures that the most informative sentiment signal, regardless of its position in the input text, dominates the final feature representation. As a result, the output vector A contains the most essential sentiment-related *n-gram* pattern found in the student's post.

3.3 Semantic Similarity Analysis (BiLSTM-RoBERTa)

To determine whether a lecturer's response adequately addresses a student's post, TWIN-Net applies a semantic similarity module. This mechanism captures the degree of alignment in meaning between the two texts, which strongly influences student satisfaction. When a reply closely reflects the student's concern, it is more likely to be seen as helpful, relevant, and supportive. The semantic similarity branch begins by concatenating the contextual embeddings of the student and lecturer's feedback. These embeddings, generated by RoBERTa, are then processed by a BiLSTM network:

$$H_{bi} = BiLSTM\,([\Phi(\mathrm{s}),\, \Phi(\mathrm{L})]) \tag{5}$$

The BiLSTM processes the input in both forward and backward directions, enabling the model to capture semantic dependencies that may appear earlier or later in the sequence. As a result, the network produces a sequence of hidden states:

$$H_{bi} = \{H_{bi,1}, H_{bi,2}, \ldots, H_{bi,T}, \} \tag{6}$$

where each hidden state $H_{bi,t}$ encodes semantic information derived from its surrounding context. This bidirectional encoding is particularly advantageous in asynchronous educational discussions, where key semantic relationships, such as references to assignment instructions or conceptual clarifications, may occur at different positions across the student and lecturer's utterances. Recognizing that not all tokens contribute equally to semantic correspondence, an attention mechanism is applied over the sequence of BiLSTM hidden states. The attention layer assigns an importance weight α_t to each hidden state and computes a weighted sum to produce a context vector that emphasizes the most informative components of the interaction:

$$v = \sum_{t=1}^{T} \alpha_t H_{bi}, t \tag{7}$$

Once the attended representations for the student's post v_S and lecturer's reply v_L have been derived, the semantic similarity between the two is computed using cosine similarity:

$$B = \frac{vs.vl}{||vs||||vl||} \tag{8}$$

where v_S and v_L represent the attended representations of the student's and the lecturer's sentences. Cosine similarity measures the angular distance between the two vectors and yields a score between -1 and 1, with higher values indicating stronger semantic alignment. In addition to being used as a continuous feature within TWIN-Net, the similarity score is also categorized into three interpretive levels: ≥ 0.75 (indicated as high), 0.60–0.75 (indicated as moderate), and < 0.60 (indicated as low), adapted from a previous study [5]. By introducing these categories, the analysis becomes more interpretable, and they are employed throughout the evaluation stage.

3.4 Emotion Feature Extraction (CNN-RoBERTa)

To enhance the semantic and sentiment dimensions, TWIN-Net integrates an emotion-extraction module that detects affective cues in students' textual output. Since emotional signals in educational discourse often manifest through short, localised expressions, such as confusion, frustration, or relief, convolutional operations are particularly effective for capturing these subtleties. Consequently, a CNN is applied to the contextual embedding sequence $\Phi(S)$ to generate fine-grained emotional representations C:

$$C = \max_{i} \sigma\left(W_{emo} * \Phi(S)_{i:i+k-1} + b_{emo}\right) \tag{9}$$

Here, W_{emo} denotes the convolution filter, k specifies the kernel size, and $\sigma(\cdot)$ represents the ReLU activation. By sliding over short token spans, the CNN isolates emotionally indicative n-gram patterns, while the subsequent *max-pooling* operation selects the most dominant activation. The resulting vector C captures the primary emotional signals present in the student's post and serves as an essential feature for assessing satisfaction within the TWIN-Net framework.

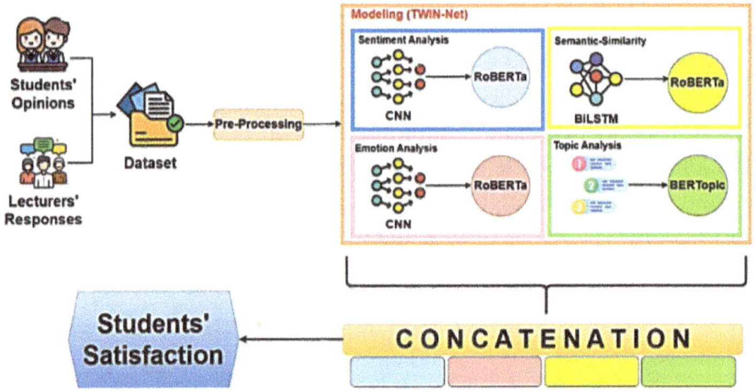

Fig. 1. The proposed architecture of the TWIN-Net model

3.5 Topic Representation (BERTopic)

To incorporate thematic context from the lecturer's responses, TWIN-Net employs BERTopic to generate topic-level representations. Using the lecturer's embedding sequence $\Phi(L)$, BERTopic clusters semantically similar utterances and produces a representative topic vector for each cluster, formalized as:

$$D = TopicVector(Cluster(\Phi(\mathrm{L}))) \tag{10}$$

The resulting topic vector D provides a concise summary of the main instructional theme present in the lecturer's reply and complements the other feature branches within TWIN-Net. Additionally, BERTopic was configured using the *paraphrase-multilingual-MiniLM-L12-v2* embedding model with min_topic_size = 10 and n_gram_range = (1,3) to capture both unigram and multi-word expressions. Dimensionality reduction and clustering were performed using the default UMAP and HDBSCAN implementations within BERTopic, with the minimum cluster size aligned to 10. Topic representations were generated using the KeyBERTInspired representation model (diversity = 0.3), and each topic was characterized by the top 10 class-based TF-IDF (c-TF-IDF) keywords. For interpretability purposes, the topic modeling process was constrained to three dominant clusters.

3.6 Feature Fusion and Multi-Class Output

After the individual feature branches generate their respective representations, TWIN-Net integrates them into a single fused vector. The sentiment feature A, semantic similarity feature B, emotion feature C, and topic representation D are concatenated to form the final unified representation:

$$TWIN - Net = A \oplus B \oplus C \oplus D \tag{11}$$

This fused vector encapsulates complementary linguistic, semantic, emotional, and thematic information extracted from student-lecturer interactions. To obtain the final satisfaction prediction, the fused representation is then transformed by an output module using Softmax to generate class probabilities. The predicted probability for each class k is given by:

$$\hat{y}_k = \frac{exp(TWIN - Net_k)}{\sum_{j=1}^{K} exp(TWIN - Net_j)} \tag{12}$$

3.7 Training Objective with 10-Fold Cross-Validation

To obtain reliable estimates of the model's performance, TWIN-Net is trained using a 10-fold cross-validation scheme. In this procedure, the dataset is partitioned into ten equal subsets. In each iteration, nine subsets are used for training, while the remaining subset serves as the test set. This procedure is repeated ten times, ensuring that every subset is used exactly once as the evaluation set. Averaging the results across all folds provides a robust estimate of model generalization and mitigates overfitting by exposing the model to multiple independent train and test splits. Within each training fold, model parameters are optimized using the categorical cross-entropy loss function $\mathcal{L}$, defined as:

$$\mathcal{L} = -\sum_{k=1}^{K} y_k log(\hat{y}_k) \tag{13}$$

4 Results and Discussion

This section presents the experimental results and discussion of the proposed TWIN-Net model. The analysis is conducted using data collected from asynchronous online discussion forums involving lecturers and students, focusing on evaluating model performance and interpreting the findings in relation to baseline approaches. The results are achieved by averaging metrics across all folds.

The dataset consists of 15,000 entries sourced from [5], encompassing various academic disciplines and gender compositions. It includes contributions from eight study programs, including Accounting (notated as Ac), Broadcasting (notated as Bc), Computer Science (notated as CS), Engineering (notated as En), Management (notated as Mn), Psychology (notated as Ps), Public Relations (notated as PR), and Visual Communication (notated as VC), with a relatively balanced distribution across fields. Male participants represent approximately 59.7% of the sample, while female participants account for 40.3%, ensuring demographic diversity and representational balance. Each discussion entry includes a satisfaction label reflecting the student's perceived level of satisfaction with the lecturer's response. These labels were collected using a structured self-report feature integrated into the discussion platform, where students were required to select one of three predefined categories (positive, neutral, or negative) at the time of submission. The three categories were explicitly presented as indicators of satisfaction level. Thus, the labeling relied on participants' self-assessment rather than manual annotation or automated classification. Since the labels were self-declared, inter-annotator agreement metrics were not applicable in this context. Entries with conflicting or inconsistent responses were removed during preprocessing, resulting in a final dataset of 15,000 valid samples. The classes are relatively well balanced, enabling fair evaluation during cross-validation.

4.1 Addressing RQ1: Does TWIN-Net Demonstrate Significant Improvement Over Previous Models for Assessing Students' Satisfaction Through the Integration of Sentiment, Emotion, Semantic Similarity, and Topic Features?

The purpose of RQ1 is to evaluate whether TWIN-Net provides a significant performance benefit over current methods for assessing students' satisfaction in asynchronous e-learning discussions. Previous approaches based on CNN, LSTM, BiLSTM, and CNN-LSTM usually focus on individual analytical components, such as sentiment, emotion, semantic similarity, or topic analysis, without integrating them into a single framework.

4.1.1 Experimental Results

Across all feature extraction methods, TWIN-Net consistently demonstrates the highest performance. In the Dataset-1, using GloVe, the model reaches an average F1-score of about 72.9%, while baseline models remain within the 72.4%-72.9% range. With fastText, performance improves noticeably, and TWIN-Net achieves 84.1%, outperforming E-SATNet by approximately 1.4%. Under BERT, TWIN-Net increases further to 87.1%, exceeding E-SATNet ($\approx$84.6%) by about 2.5%. The best results arise with RoBERTa, where TWIN-Net reaches a macro-F1 of 92.72%, surpassing E-SATNet ($\approx$88.62%) by more than 4%. These results highlight the advantages of richer contextual embeddings and demonstrate how TWIN-Net's multi-branch architecture effectively captures emotional cues across varied linguistic contexts. Moreover, in Dataset-2, TWIN-Net consistently achieves the highest macro-F1 across all embeddings, improving from ($\approx$71.1%) with GloVe to ($\approx$89.0%) with RoBERTa, while maintaining a clear performance advantage over baseline models and E-SATNet.

4.2 Addressing RQ2: Does the Proposed TWIN-Net Sustain Stable Performance Across Different Various Embedding Techniques Including GloVe, FastText, BERT, and RoBERTa?

The examination of RQ2 focuses on evaluating whether TWIN-Net maintains stable predictive performance across different embedding techniques, specifically GloVe, fastText, BERT, and RoBERTa. Each embedding approach offers different levels of contextual understanding and semantic detail, which can affect the performance of deep learning models. Models that depend heavily on a particular embedding type often show inconsistent results as the underlying text representations vary, indicating limited robustness and poor generalization.

4.2.1 Ablation Study

To evaluate the contribution of each architectural branch, an ablation study was performed using RoBERTa embeddings, under which TWIN-Net attained the highest overall accuracy, as shown in Table 2. The findings demonstrate that removing any branch consistently degrades macro-F1 performance, underscoring the complementary nature of the modules. Excluding the topic branch (CNN + LSTM) reduces performance by 4.74%, confirming the relevance of thematic cues from lecturer responses (Table 1).

Table 1. Students' emotions results based on several feature extraction techniques.

Features	Classifier	Happiness			Anger			Fear			Disgust			Sadness		
		Pre	Rec	F1	Pre	Rec	F1	Pre	Rec	F1	Pre	Rec	F1	Pre	Rec	F1
GloVe	CNN	72,34	72,56	72,41	72,12	72,44	72,27	72,69	72,35	72,48	72,18	72,53	72,3	72,59	72,25	72,42
	LSTM	72,45	72,78	72,61	72,23	72,49	72,36	72,91	72,67	72,79	72,4	72,72	72,54	72,63	72,4	72,51
	BiLSTM	72,58	72,9	72,74	72,35	72,66	72,5	72,85	72,52	72,68	72,5	72,84	72,67	72,71	72,49	72,6
	CNN + LSTM	72,71	72,94	72,82	72,48	72,77	72,63	72,98	72,71	72,84	72,61	72,89	72,75	72,79	72,57	72,68
	E-SATNet	72,83	73,02	72,92	72,59	72,81	72,7	73	72,76	72,88	72,72	72,95	72,83	72,86	72,63	72,75
	TWIN-Net	72,95	73,15	73,05	72,71	72,91	72,81	73,12	72,88	73	72,84	73,03	72,94	72,97	72,74	72,85
fastText	CNN	75,76	75,95	75,83	75,44	75,89	75,61	76,1	75,78	75,92	75,63	75,98	75,79	75,96	75,65	75,84
	LSTM	77,29	77,54	77,41	77,02	77,3	77,16	77,6	77,31	77,44	77,18	77,47	77,33	77,43	77,15	77,29
	BiLSTM	78,61	78,82	78,71	78,34	78,59	78,46	78,92	78,61	78,75	78,48	78,73	78,6	78,75	78,46	78,61
	CNN + LSTM	80,02	80,24	80,13	79,73	79,99	79,86	80,33	80,02	80,16	79,88	80,12	80	80,13	79,86	80
	E-SATNet	82,44	82,65	82,54	83,18	83,43	83,31	83,7	82,43	82,57	82,27	82,52	82,4	82,52	82,26	82,39
	TWIN-Net	83,87	84,06	83,96	84,6	84,85	84,73	85,13	84,84	83,99	83,7	83,93	83,82	83,94	83,69	83,81
BERT	CNN	78,22	78,45	78,33	77,89	78,22	78,05	78,48	78,16	78,29	78,01	78,36	78,21	78,17	78,04	78,09
	LSTM	79,9	80,13	80,01	79,57	79,9	79,73	80,16	79,84	79,97	79,69	80,04	79,89	79,85	79,72	79,77
	BiLSTM	81,29	81,52	81,4	80,96	81,29	81,12	81,55	81,23	81,36	81,08	81,43	81,28	81,24	81,11	81,16
	CNN + LSTM	82,73	82,96	82,84	82,4	82,73	82,56	82,99	82,67	82,8	82,52	82,87	82,72	82,68	82,55	82,6
	E-SATNet	84,17	84,4	84,28	85,84	85,17	85	84,43	84,78	84,74	84,96	84,31	84,16	85,12	85,99	85,04
	TWIN-Net	86,61	86,84	86,72	87,28	87,61	87,44	86,87	86,85	86,88	87,14	87,15	87,16	87,26	87,23	87,38
RoBERTa	CNN	81,66	81,93	81,78	81,33	81,61	81,44	81,92	81,59	81,72	81,46	81,82	81,66	81,6	81,46	81,54
	LSTM	84,46	84,73	84,58	84,13	84,41	84,24	84,72	84,39	84,52	84,26	84,62	84,46	84,4	84,26	84,34
	BiLSTM	87,51	87,78	87,63	87,18	87,46	87,29	87,77	87,44	87,57	87,31	87,67	87,51	87,45	87,31	87,39
	CNN + LSTM	88,01	88,28	88,13	87,68	87,96	87,79	88,27	87,94	88,07	87,81	88,17	88,01	87,95	87,81	87,89
	E-SATNet	90,63	90,49	90,57	88,49	88,85	88,69	88,95	87,62	87,75	87,36	87,64	87,47	88,48	88,75	88,6
	TWIN-Net	93,79	93,65	93,73	92,65	92,01	92,85	92,11	92,78	92,91	92,52	92,8	92,63	91,34	91,61	91,46

Removing the BiLSTM branch decreases performance by 5.24%, the emotion branch by 8.29%, and the sentiment CNN branch by 10.94%, the largest drop confirming that emotional and sentiment features are critical, alongside semantic alignment, for accurate prediction.

Table 2. The performance results observed following ablation in machine learning models.

Variant	Active Components	Macro-F1 (%)	Δ vs Full TWIN-Net
Full TWIN-Net	Sentiment + Emotion + Semantic Similarity + Topic	92.72	-
w/o Multi-Branch (E-SATNet)	Sentiment + Emotion + Similarity	90.42	-2.30
w/o Topic Branch (CNN + LSTM)	Sentiment + Emotion	87.98	-4.74
w/o Semantic Similarity (BiLSTM)	Emotion + Topic	87.48	-5.24
w/o Emotion Branch (LSTM)	Sentiment + Similarity + Topic	84.43	-8.29
w/o Sentiment Branch (CNN)	Emotion + Similarity + Topic	81.78	-10.94

4.3 Statistical Significance Analysis

To examine pairwise differences, the Wilcoxon signed-rank test was applied to compare TWIN-Net with each baseline, as recommended for non-parametric paired comparisons in machine learning studies [19]. As shown in Table 3, all comparisons were statistically significant at $\alpha = 0.05$ ($p = 0.00060$–0.00065), with $W = 0$ and $Z = 3.408$, indicating consistent directional improvements. The large effect size ($r_{rb} = 0.88$) confirms very strong practical significance, consistent with established interpretation guidelines [20, 21].

Table 3. The Results of Wilcoxon signed-rank test.

Comparison	p-value	W	Z	Mean	Standard Deviation	Effect Size (r_{rb})	Significant (α = 0.05)
TWIN-Net vs CNN	0.00065	0.0	3.408	10.95	0.73	0.88	Yes
TWIN-Net vs LSTM	0.00065	0.0	3.408	8.15	0.73	0.88	Yes
TWIN-Net vs BiLSTM	0.00065	0.0	3.408	5.10	0.73	0.88	Yes

(*continued*)

Table 3. (*continued*)

Comparison	p-value	W	Z	Mean	Standard Deviation	Effect Size (r_{rb})	Significant (α = 0.05)
TWIN-Net vs CNN + LSTM	0.00065	0.0	3.408	4.60	0.73	0.88	Yes
TWIN-Net vs E-SATNet	0.00060	0.0	3.408	3.90	1.00	0.88	Yes

Figure 2 illustrates TWIN-Net's paired improvements over each baseline across the fifteen emotion-metric pairs. The violin and box plots reveal that all gains are strictly positive, aligning with the Wilcoxon result (W = 0) and confirming uniformly positive, directionally consistent improvements. The relatively narrow interquartile ranges indicate stable performance gains. Larger mean improvements are observed against simpler models (CNN, LSTM), while smaller yet consistent gains over stronger hybrids (CNN + LSTM, E-SATNet) highlight sustained superiority.

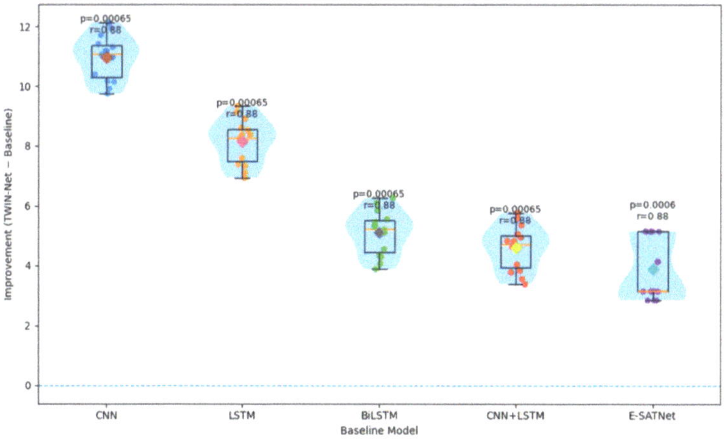

Fig. 2. Paired performance improvements of TWIN-Net over baseline models across fifteen emotion–metric pairs.

4.4 Discussion

The valence-arousal distribution in Fig. 3 shows that the three topic clusters, *Lecturer Responses, Learning Discussion*, and *Learning Model* occupy distinct affective regions. To formalize the emotional patterns within each topic, we define the set of valence–arousal points for the topic t as $\mathcal{V}_t = \{(v_i, a_i)\}$, and summarize each topic's emotional profile using its centroid $\mu_t = (\overline{v}_t, \overline{a}_t)$. The Lecturer Responses cluster exhibits the highest arousal values, indicating more directive, evaluative, or corrective communication from instructors. In contrast, the Learning Model cluster lies predominantly in low-arousal regions, reflecting calmer, reflective, and explanatory content. The Learning Discussion ellipse lies between these two extremes, functioning as an affective bridge that blends questioning, meaning negotiation, and collaborative understanding. The differences between centroids $\boldsymbol{\mu}_t$ reveal that topic categories shape emotional tone as much as the underlying emotion labels themselves. High-arousal lecturer feedback shifts the distribution upward, while conceptual modeling tasks draw responses downward toward low-arousal reflective states.

Meanwhile, learning discussions occupy a balanced mid-valence, mid-arousal region, suggesting that affective expression intensifies or relaxes depending on the instructional role. These findings indicate that emotional patterns in asynchronous learning are strongly structured by topical context: the type of activity, including responding, discussing, or modeling, produces distinct affective signatures that can be quantified through their valence arousal distributions. Building on this affective structure, the distribution of high-intensity emotional responses across academic programs further clarifies how these topic-driven dynamics manifest at the group level.

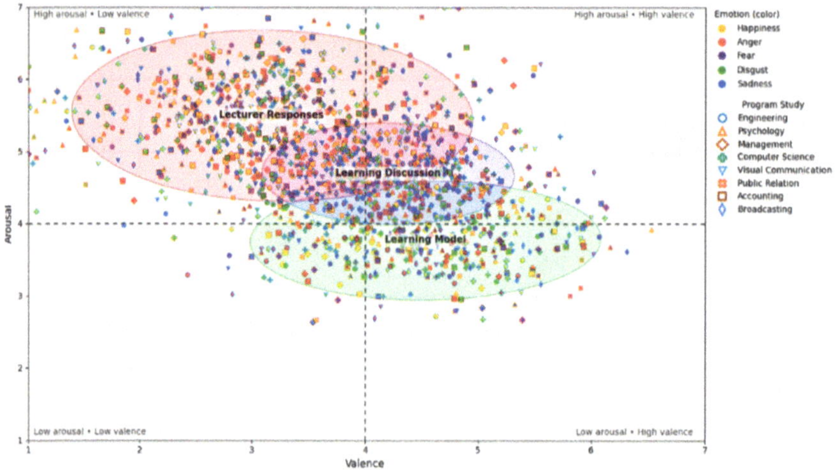

Fig. 3. Student emotion distribution in the valence-arousal space across academic programs. Colors indicate emotion categories, shapes denote programs, and ellipses mark the three dominant clusters.

The emotional distribution in Fig. 4 shows that negative effect is heavily concentrated in several academic groups, especially in Psychology and communication-focused programs. The highest proportion of high-intensity negative emotions ($\geq$0.75) appears among female Psychology students in the Learning Model topic (42.34%, or 105 out of 248 responses). A similar trend is seen among male Psychology students during E-Learning Discussion, with 41.10% (104 out of 253) displaying severe negative emotions. Communication-oriented majors follow comparable patterns: female Public Relations students in Learning Model sessions reach 42.65% (87 out of 204), while female Visual Communication students in E-learning Discussion record 37.50% (42 out of 112). These elevated percentages suggest that students in these programs may experience greater emotional vulnerability during interpretive, discussion-based, or feedback-driven tasks, possibly due to higher cognitive-emotional demands or misalignment between instructional clarity and learner expectations.

These elevated percentages suggest that students in these programs may experience greater emotional vulnerability during interpretive, discussion-based, or feedback-driven tasks, possibly due to higher cognitive-emotional demands or misalignment between instructional clarity and learner expectations. In summary, although TWIN-Net demonstrates strong text-based performance, it still lacks multimodal inputs such as speech prosody, facial expressions, or gestures [22]. These non-verbal cues carry essential emotional and behavioral information that text alone cannot capture, meaning the model may be less effective when learners express meaning or emotion primarily through vocal or visual channels. From a practical standpoint, the proposed architecture nevertheless offers strong potential for integration into learning management systems to support automated monitoring of student satisfaction and assist instructors in identifying engagement patterns and response effectiveness. Such deployment could enable timely pedagogical adjustments and more data-driven instructional strategies. However, scalability and real-time operational integration across diverse educational environments require further empirical investigation.

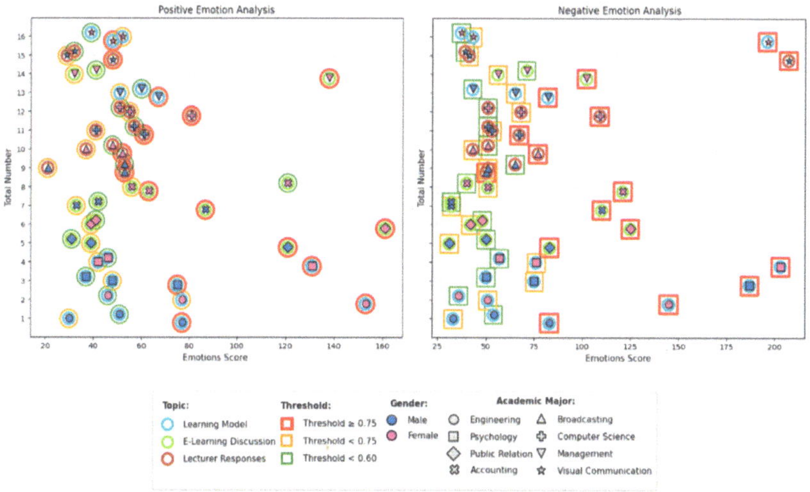

Fig. 4. Positive and negative emotion distributions in relation to semantic similarity and topic categories. The x-axis represents emotion scores and the y-axis indicates total frequency; colors denote topics and outlined markers indicate similarity thresholds.

5 Conclusion

This study introduced TWIN-Net, a deep learning framework that integrates sentiment, semantic similarity, emotion, and topic analysis tomeasure students' satisfaction in asynchronous online discussions. By combining CNN-RoBERTa, BiLSTM-RoBERTa, and BERTopic within a wide multi-branch integration architecture, the model effectively captures emotional cues, semantic coherence, and thematic alignment, achieving a 92.72% F1-score on a 15,000-entry dataset across eight academic programs.

Although the framework demonstrates notable strengths, certain limitations persist. First, it relies solely on textual input and does not incorporate multimodal signals, such as speech prosody, facial expressions, or gestures, which could provide richer insights into affective and behavioral states. Incorporating audio-visual modalities may strengthen the detection of subtle emotional states. Second, the dataset is drawn from a single institutional context, which may restrict generalizability to other educational environments with distinct communication practices. Third, while BERTopic and similarity modeling yield interpretable structures, deeper neural layers (e.g., RoBERTa) remain partially opaque in their internal representations. Future research may pursue multimodal integration, cross-institutional validation, additional validation mechanisms, and attention-based explainability approaches to further enhance robustness, interpretability, and applicability in authentic educational settings.

Acknowledgments. The authors acknowledge the financial assistance from Universitas Indonesia (UI) via the PUTI program, as stated in Agreement No. PKS-318/UN2.RST/HKP.05.00/2025.

References

1. Sandiwarno, S., Sensuse, D.I., Santoso, H.B., Hidayat, D.S., Nyamawe, A.S.: SEMAR: a multi-task term weighting approach for sentiment and emotion-based e-learning users' satisfaction analysis. IEEE Access **13**, 187584–187601 (2025). https://doi.org/10.1109/ACCESS.2025.3627657
2. Sandiwarno, S., Sensuse, D.I., Santoso, H.B., Hidayat, D. S., Mishbah, M.: SENSEI: a novel architecture of deep multi-task neural network for analyzing e-learning lecturers' and students' satisfaction. In: 2024 Ninth International Conference on Informatics and Computing (ICIC), pp 1–7. IEEE, Medan (2024)
3. Vedavathi, N., Anil, K.K.M.: E-learning course recommendation based on sentiment analysis using hybrid Elman similarity. Knowl.-Based Syst. **259** (2023). https://doi.org/10.1016/j.knosys.2022.110086
4. Hasnine, M.N., Bui, H.T.T., Tran, T.T.T., Nguyen, H.T., Akçapınar, G., Ueda, H.: Students' emotion extraction and visualization for engagement detection in online learning. Procedia Comput. Sci. **192**, 3423–3431 (2021)
5. Sandiwarno, S., Sensuse, D.I., Santoso, H.B., Hidayat, D.S., Nyamawe, A.S., Yousif. A.: E-SATNet: evaluating student satisfaction with lecturer responses in asynchronous online discussions using sentiment and semantic similarity analysis. Big Data Cognitive Comput. **9**(9), 228 (2025). https://doi.org/10.3390/bdcc9090228
6. Chen, C. M., Li, M. C., Chang, W. C., Chen, X. X.: Developing a topic analysis instant feedback system to facilitate asynchronous online discussion effectiveness. Comput. Educ. **163** (2021). https://doi.org/10.1016/j.compedu.2020.104095
7. Lin, E., Sun, J., Chen, H., Mahoor, M.H.: Data quality matters: suicide intention detection on social media posts using RoBERTa-CNN. In: Proceedings of the Annual International Conference of the IEEE Engineering in Medicine and Biology Society (EMBS), pp. 1–5. Orlando (2024)
8. Rahman, Md.M., Shiplu, A.I., Watanobe, Y., Alam, Md.A.: RoBERTa-BiLSTM: a context-aware hybrid model for sentiment analysis (2024)
9. Wang, Z., Chen, J., Chen, J., Chen, H.: Identifying interdisciplinary topics and their evolution based on BERTopic. Scientometrics **129**, 7359–7384 (2024). https://doi.org/10.1007/s11192-023-04776-5

10. Ekman, P.: An argument for basic emotions. Cogn. Emot. **6**(3–4), 169–200 (1992). https://doi.org/10.1080/02699939208411068
11. Liu, S., Liu, S., Liu, Z., Peng, X., Yang, Z.: Automated detection of emotional and cognitive engagement in MOOC discussions to predict learning achievement. Comput. Educ. **181** (2022). https://doi.org/10.1016/j.compedu.2022.104461
12. Koufakou, A.: Deep learning for opinion mining and topic classification of course reviews. Educ. Inf. Technol. **29**, 2973–2997 (2024). https://doi.org/10.1007/s10639-023-11736-2
13. Paneru, B., Thapa, B., Paneru, B.: Sentiment analysis of movie reviews: a flask application using CNN with RoBERTa embeddings. Syst. Soft Comput. **7** (2025). https://doi.org/10.1016/j.sasc.2025.200192
14. Semary, N.A., Ahmed, W., Amin, K., Pławiak, P., Hammad, M.: Improving sentiment classification using a RoBERTa-based hybrid model. Front. Human Neurosci. **17** (2023). https://doi.org/10.3389/fnhum.2023.1292010
15. Putra, R. P., Setiawan, E. B.: RoBERTa as semantic approach for big five personality prediction using artificial neural network on twitter. In: 2022 International Conference on Advanced Creative Networks and Intelligent Systems (ICACNIS), Bandung, Indonesia (2022)
16. Lak, A.J., Boostani, R., Alenizi, F.A., Mohammed, A.S., Fakhrahmad, S.M.: RoBERTa, ResNeXt and BiLSTM with self-attention: the ultimate trio for customer sentiment analysis. Appl. Soft Comput. **164** (2024). https://doi.org/10.1016/j.asoc.2024.112018
17. Bakhti, K., Niu, Z., Nyamawe, A. S.: A new scheme for citation classification based on convolutional neural networks. In: Proceedings of the International Conference on Software Engineering and Knowledge Engineering, SEKE (2018)
18. Bakhti, K., Niu, Z., Nyamawe, A. S.: Semi-automatic annotation for citation function classification. In: Proceedings - 2018 International Conference on Control, Artificial Intelligence, Robotics & Optimization (ICCAIRO). IEEE, Prague (2018)
19. Demšar, J.: Statistical comparisons of classifiers over multiple data sets. J. Mach. Learn. Res. **7**, 1–30 (2006)
20. Tomczak, M., Tomczak, E.: The need to report effect size estimates revisited. an overview of some recommended measures of effect size. Trends Sport Sci. **1**(21), 19–25 (2014)
21. Kerby, D.S.: The simple difference formula: an approach to teaching nonparametric correlation. Compre. Psychol. **3** (2014). https://doi.org/10.2466/11.it.3.1
22. Ani, N., Ali, N.M., Ayumi, V: Human gesture recognition for elderly people using user training interaction data. In: Lecture Notes in Computer Science (including subseries Lecture Notes in Artificial Intelligence and Lecture Notes in Bioinformatics), pp. 109–118 (2024)

On PyTorch Bilinear Neural Network-Based Prediction for Knee Osteoarthritis

Nguyen Minh Tuan[1], Phayung Meesad[2], Nguyen Hong Son[1], and Maleerat Maliyaem[3](✉)

[1] Faculty of Information Technology, Posts and Telecommunications Institute of Technology, 11 Nguyen Dinh Chieu, Sai Gon ward, Ho Chi Minh 700000, Viet Nam
{minhtuan,sonngh}@ptit.edu.vn

[2] Department of Information Technology Management, King Mongkut's University of Technology North Bangkok, Bangkok, Thailand
phayung.m@itd.kmutnb.ac.th

[3] Department of Information Technology, King Mongkut's University of Technology North Bangkok, Bangkok, Thailand
maleerat.m@itd.kmutnb.ac.th

Abstract. Knee osteoarthritis (KOA) is a prevalent degenerative joint disease that significantly affects mobility and quality of life, particularly among aging populations. Early and accurate prediction of disease severity is essential for effective clinical decision-making and personalized treatment planning. This study proposes a PyTorch-based bilinear neural network framework for predicting Kellgren–Lawrence (KL) grades of knee osteoarthritis using structured clinical and biochemical data. The proposed model integrates a recurrent neural network (RNN) feature extractor with a bilinear interaction layer to capture higher-order relationships among heterogeneous patient features. The bilinear component enables the learning of multiplicative dependencies between latent representations, enhancing the model's ability to discriminate between closely related disease stages. To address real-world data challenges, the framework incorporates automated data normalization, label remapping, and robust handling of missing and non-standard numeric formats. The model is trained and evaluated on a curated clinical dataset, with performance assessed using accuracy, precision, recall, and F1-score computed per epoch. Experimental results demonstrate that the proposed bilinear neural network consistently outperforms baseline RNN architectures, achieving improved classification stability and superior predictive performance across all KL grade categories. The attainment provides the potential of bilinear neural architectures implemented in PyTorch as a reliable and scalable solution for clinical decision support in knee osteoarthritis severity prediction.

Keywords: Knee Osteoarthritis · Bilinear Neural Networks · PyTorch · KL Grade · Clinical Prediction

S. Sodsee et al. (Eds.): IC2IT 2026, LNNS 5590, pp. 26–34, 2026.
https://doi.org/10.1007/978-3-032-28048-0_3

1 Introduction

Knee osteoarthritis (KOA) is a progressive musculoskeletal disorder characterized by cartilage degeneration, joint space narrowing, and chronic pain, which significantly impacts mobility and quality of life, particularly among aging populations. Accurate assessment of disease severity, commonly performed using the Kellgren–Lawrence (KL) grading system, plays a critical role in guiding clinical decision-making and treatment planning. However, traditional radiographic interpretation is time-consuming and subject to inter-observer variability, motivating the development of automated and data-driven diagnostic frameworks. Recent advances in deep learning have demonstrated strong potential for modeling complex, high-dimensional biomedical data using bilinear and convolutional neural network architectures. Mastouri *et al.* [1] introduced a bilinear convolutional neural network for lung nodule classification from computed tomography images, showing that bilinear feature interactions can significantly enhance fine-grained visual discrimination. Similarly, Dai *et al.* [2] proposed a bilinear CNN framework based on subcategory-similarity measurements, achieving improved performance in fine-grained classification tasks by explicitly modeling pairwise feature relationships.

Table 1. Summary of demographic, clinical, and radiographic characteristics of the dataset

Characteristic	Value	Percentage (%)
Number of subjects	160	100.0
Sex		
Male	34	21.2
Female	126	78.8
Age (years)	64.2 ± 8.7	–
BMI (kg/m^2)	25.0	–
Kellgren–Lawrence grade		
Grade 2	56	35.0
Grade 3	89	55.6
Grade 4	15	9.4

Beyond medical imaging, bilinear and graph-based neural models have also been applied in complex classification domains. Zhang *et al.* [3] developed a bilinear graph neural network to enhance web service classification, demonstrating the effectiveness of bilinear feature fusion in capturing structural dependencies within high-dimensional data. In the context of dermatological applications, Ahmad *et al.* [4] and Calderón *et al.* [5] proposed bilinear convolutional frameworks for skin disease and lesion classification, respectively, reporting substantial improvements in classification robustness and sensitivity across multiple disease categories. Recent work has further explored integrating bilinear mechanisms with attention and feature fusion strategies. Fang *et al.* [6] introduced a lightweight bilinear convolutional network incorporating attention mechanisms for agricultural disease classification, highlighting the capacity of bilinear models to maintain high accuracy while reducing computational complexity. In biomedical diagnosis, Liu *et al.* [7] demonstrated that bilinear convolutional neural networks can effectively capture subtle inter-class variations in fine-grained breast cancer classification, reinforcing the value of higher-order feature interactions in clinical prediction tasks.

In parallel, bilinear neural network formulations have also been investigated in theoretical and recurrent learning frameworks. Tuan and Meesad [8,9] proposed bilinear neural and recurrent neural architectures for solving nonlinear and fractional differential equations, illustrating the mathematical robustness and representational power of bilinear models in complex dynamical systems. Despite these advancements, the application of bilinear neural networks to structured clinical data for knee osteoarthritis severity prediction remains relatively underexplored. Most existing approaches focus primarily on convolutional architectures applied to radiographic images, with limited attention to modeling higher-order interactions among heterogeneous patient attributes such as demographics, biochemical markers, and clinical symptoms.

To address this gap, this study proposes a PyTorch-based bilinear neural network framework that integrates recurrent feature modeling with bilinear interaction learning for automated prediction of Kellgren-Lawrence grades. By explicitly capturing pairwise dependencies among sequential clinical features, the proposed approach aims to improve class separability and generalization performance in multi-class KOA severity classification. Experimental results demonstrate that the proposed model achieves high predictive accuracy and robust performance, highlighting its potential as a reliable clinical decision support tool for knee osteoarthritis assessment. Table 1 summarizes the demographic, clinical, and radiographic characteristics of the study population. The dataset includes 160 subjects with knee osteoarthritis, with a predominance of female participants (78.8%). The mean age was 64.2 ± 8.7 years, and the average body mass index was $25.0\,\text{kg/m}^2$. Radiographic assessment using the Kellgren–Lawrence grading system showed that most subjects were classified as grade 3 (55.6%), followed by grade 2 (35.0%) and grade 4 (9.4%). Overall, the table indicates a cohort mainly composed of patients with moderate to severe osteoarthritis.

2 Methodology

This section presents the proposed PyTorch-based bilinear neural network framework for predicting Kellgren–Lawrence (KL) grades of knee osteoarthritis. The methodology includes dataset preprocessing, feature normalization, label encoding, model architecture, training strategy, and evaluation protocol (shown in Fig. 1).

2.1 Dataset Description

The dataset consists of structured clinical, anthropometric, and biochemical attributes collected from patients undergoing assessment for knee osteoarthritis. The feature set includes demographic variables (e.g., age, sex), body composition indicators (e.g., body mass index, waist circumference), clinical history factors, medication usage records, laboratory test results, and symptom-based scores such as the Western Ontario and McMaster Universities Osteoarthritis Index (WOMAC). The target variable is the Kellgren–Lawrence (KL) grade, which categorizes disease severity into discrete ordinal classes based on radiographic evaluation.

2.2 Data Preprocessing

A standardized preprocessing pipeline is applied to the raw dataset to ensure numerical consistency and robustness. Non-standard numeric formats, including decimal values represented using commas, are automatically converted into floating-point format. All relevant features are cast to numeric types, while non-informative identifier columns are removed.

Missing values are imputed using the median to preserve the underlying feature distributions and mitigate the influence of outliers. The target KL grade labels are remapped to a contiguous integer range from 0 to $C - 1$, where C denotes the number of unique classes. This encoding ensures compatibility with the categorical cross-entropy loss function used during model optimization.

All features are normalized using z-score standardization, defined as:

$$x' = \frac{x - \mu}{\sigma}, \tag{1}$$

where μ and σ represent the mean and standard deviation of each feature, respectively.

2.3 Train–Test Partitioning

The dataset is partitioned into training and testing subsets using stratified sampling to preserve the distribution of KL grade classes across both sets. A fixed random seed is employed to ensure reproducibility. The training set is used for parameter optimization, while the testing set serves as an independent validation set for assessing generalization performance.

2.4 Model Architecture

The proposed architecture comprises three primary components: an RNN-based feature extractor, a bilinear interaction layer, and a classification head.

RNN Feature Extraction. Each patient record is represented as a one-dimensional sequence of normalized feature values, which is processed by a multi-layer recurrent neural network (RNN). The RNN learns latent feature representations that capture inter-feature dependencies. The final hidden state is extracted as a compact embedding vector for each input sample.

Formally, given an input sequence $\mathbf{X} = \{x_1, x_2, \ldots, x_T\}$, the RNN computes a hidden representation $\mathbf{h}_T$ at the final time step, which serves as the learned feature embedding.

Bilinear Interaction Layer. To model higher-order feature dependencies, a bilinear interaction layer is applied to the RNN output. This layer captures pairwise multiplicative relationships between latent feature dimensions, enhancing the network's ability to distinguish between closely related disease severity classes.

Given the latent feature vector $\mathbf{h} \in \mathbb{R}^d$, the bilinear transformation is defined as:

$$\mathbf{z} = \mathbf{h}^\top \mathbf{W} \mathbf{h} + \mathbf{b}, \tag{2}$$

where $\mathbf{W} \in \mathbb{R}^{d \times d}$ is a learnable weight matrix and $\mathbf{b}$ denotes a bias vector. The resulting representation $\mathbf{z}$ encodes second-order feature interactions.

Classification Head. The bilinear output is passed through a fully connected layer with a nonlinear activation function, followed by a softmax layer that produces a probability distribution over the C KL grade classes. The predicted class corresponds to the index of the maximum posterior probability.

2.5 Training Strategy

Model training is performed using mini-batch gradient descent with the Adam optimizer. The categorical cross-entropy loss function is employed to minimize the discrepancy between predicted class probabilities and ground-truth labels. Training proceeds for a fixed number of epochs, and model parameters are updated iteratively based on the computed gradients.

To monitor convergence and detect overfitting, the model is evaluated on the validation set at the end of each epoch. Both training and validation metrics are recorded and visualized to facilitate post-training analysis.

2.6 Evaluation Metrics

Model performance is evaluated using multi-class classification metrics, including accuracy, precision, recall, and F1-score. Macro-averaging is applied to account for potential class imbalance by assigning equal importance to each KL grade category.

2.7 Implementation Details

The proposed framework is implemented using the PyTorch deep learning library. All experiments are conducted with fixed random seeds to ensure reproducibility. Data loading and batching are handled using PyTorch's `Dataset` and `DataLoader` utilities, enabling efficient mini-batch processing and GPU acceleration where available. Training and validation metrics are logged per epoch and visualized using dotted-line plots to illustrate convergence behavior and comparative performance between the training and validation phases (Table 2).

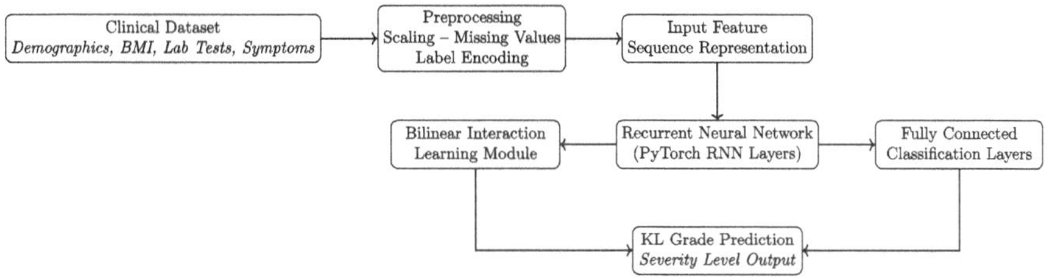

Fig. 1. A4-formatted architecture of the proposed PyTorch-based bilinear neural network for knee osteoarthritis severity prediction. The framework integrates clinical data preprocessing, recurrent feature modeling, bilinear interaction learning, and fully connected classification to predict Kellgren–Lawrence (KL) grades.

Table 2. Performance summary of the proposed PyTorch-based bilinear neural network for knee osteoarthritis severity prediction at the final training epoch.

Metrics	Value
Training Loss	0.0075
Training Accuracy	0.9900
Validation Loss	0.0648
Validation Accuracy	0.9688
Precision (Macro)	0.9825
Recall (Macro)	0.8889
F1-score (Macro)	0.9243

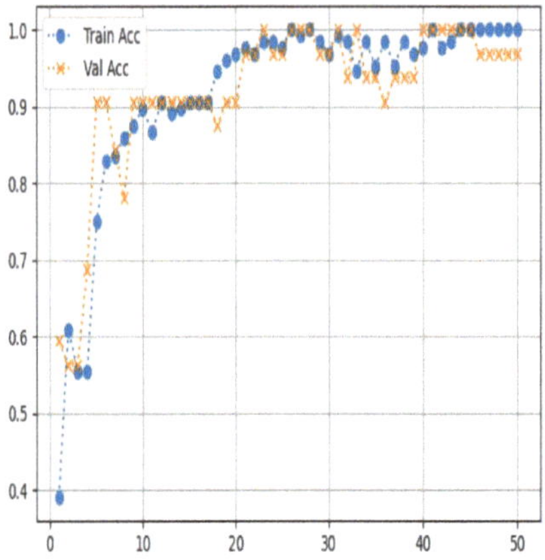

Fig. 2. Training and validation accuracy curves of the proposed bilinear neural network for knee osteoarthritis severity prediction across 50 epochs.

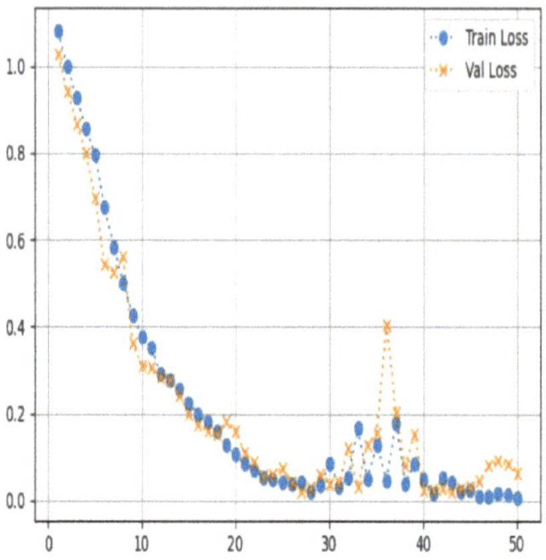

Fig. 3. Training and validation loss curves of the proposed bilinear neural network for knee osteoarthritis severity prediction across 50 epochs.

3 Numerical Results

The proposed PyTorch-based bilinear neural network was trained for 50 epochs, and its performance was evaluated using standard multi-class classification metrics, including accuracy, precision, recall, and F1-score. The evolution of training and validation loss and accuracy over epochs is illustrated in Fig. 2 and Fig. 3. At the final training epoch (Epoch 50), the model achieved a training loss of 0.0075 and a training accuracy of 0.9900, indicating near-perfect fitting on the training data. This result reflects the model's strong capacity to learn discriminative feature representations and higher-order interactions through the bilinear layer. On the validation set, the model achieved a loss of 0.0648 and an accuracy of 0.9688, demonstrating robust generalization. The relatively small gap between training and validation accuracy suggests that overfitting is effectively controlled, despite the high model expressiveness.

In terms of class-wise performance, the model achieved a macro-averaged precision of 0.9825, indicating a low false-positive rate across all Kellgren–Lawrence (KL) grade categories. The macro-averaged recall reached 0.8889, reflecting the model's strong ability to correctly identify samples from each severity class, though with a slightly higher tendency toward false negatives in certain classes. The resulting F1-score of 0.9243 highlights a well-balanced trade-off between precision and recall. The training curves show rapid, stable convergence within the first 20 epochs, with both the training and validation losses decreasing sharply while accuracy increases consistently. Minor fluctuations in validation loss and recall are observed in later epochs, which can be attributed to class imbalance and the inherent difficulty of distinguishing between adjacent KL grades. Nevertheless, the overall trend remains stable, confirming the reliability and robustness of the proposed bilinear neural network framework. These results demonstrate that the proposed model achieves high predictive accuracy and strong classification consistency, making it a promising tool for automated assessment of knee osteoarthritis severity and clinical decision support.

4 Conclusion

This study presented a PyTorch-based bilinear neural network framework for the automated prediction of knee osteoarthritis severity using structured clinical and biomechanical data. By integrating recurrent modeling with bilinear feature interactions, the proposed approach effectively captured both sequential dependencies and higher-order relationships among patient attributes, leading to strong discriminative performance across Kellgren–Lawrence (KL) grade categories. Experimental results demonstrated that the model achieved near-perfect training accuracy and strong generalization capability, with a validation accuracy of 96.88%, a macro-averaged precision of 98.25%, and an F1-score of 92.43%. The consistent convergence behavior observed in the learning curves further confirms the stability and robustness of the training process. These findings indicate that the bilinear attention mechanism significantly improves class separability, particularly for clinically adjacent KL severity levels. Despite the promising performance, certain limitations remain. The reduced recall in some classes suggests challenges in distinguishing borderline severity grades, which may be influenced by class imbalance and overlapping clinical features. Future work will focus on incorporating larger, more diverse datasets, advanced regularization strategies, and multimodal inputs, such as radiographic images and longitudinal patient records. Additionally, explainability techniques will be explored to enhance clinical interpretability and trust in real-world deployment. Overall, the proposed bilinear neural network framework offers a reliable and effective solution for predicting knee osteoarthritis severity and holds substantial potential as a clinical decision-support tool to assist healthcare professionals in early diagnosis, disease monitoring, and personalized treatment planning.

Acknowledgments. The authors would like to thank all individuals and institutions involved in the data collection and preparation processes.

Declarations

Availability of Data and Materials. The author declares no conflict of interest. The dataset is available upon request.

Funding. Not applicable.

Competing Interests. The authors confirm that there are no conflicts of interest.

References

1. Mastouri, R., Khlifa, N., Neji, H., Hantous-Zannad, S.: A bilinear convolutional neural network for lung nodules classification on CT images. Int. J. Comput. Assist. Radiol. Surg. **16**(1), 91–101 (2021). https://doi.org/10.1007/s11548-020-02283-z
2. Dai, X., Gong, S., Zhong, S., Bao, Z.: Bilinear CNN model for fine-grained classification based on subcategory-similarity measurement. Appl. Sci. **9**(2), 301 (2019). https://doi.org/10.3390/app9020301
3. Zhang, L., et al.: Bilinear graph neural network-enhanced web services classification. In: Proceedings of the IEEE International Conference on High Performance Computing and Communications, pp. 189–196 (2021). https://doi.org/10.1109/HPCC-DSS-SmartCity-DependSys53884.2021.00051
4. Ahmad, B., Usama, M., Ahmad, T., Saeed, A., Khatoon, S., Hu, L.: Bilinear-convolutional neural network using a matrix similarity-based joint loss function for skin disease classification (2024). arXiv:2406.00696. https://doi.org/10.48550/arXiv.2406.00696
5. Calderón, C., Sanchez, K., Castillo, S., Arguello, H.: BILSK: a bilinear convolutional neural network approach for skin lesion classification. Comput. Methods Prog. Biomed. Update **1**, 100036 (2021). https://doi.org/10.1016/j.cmpbup.2021.100036
6. Fang, T., Zhang, J., Qi, D., Gao, M.: BLSENet: a novel lightweight bilinear convolutional neural network based on attention mechanism and feature fusion strategy for apple leaf disease classification. J. Food Qual. **2024**, 1–11 (2024). https://doi.org/10.1155/2024/5561625
7. Liu, W., Juhas, M., Zhang, Y.: Fine-grained breast cancer classification with bilinear convolutional neural networks (BCNNs). Front. Genet. **11**, 547327 (2020). https://doi.org/10.3389/fgene.2020.547327
8. Tuan, N.M., Meesad, P.: A bilinear neural network method for solving a generalized fractional (2+1)-dimensional Konopelchenko-Dubrovsky-Kaup-Kupershmidt equation. Int. J. Theor. Phys. **64**, 17 (2025)
9. Tuan, N.M., Meesad, P.: Bilinear recurrent neural network for a modified Benney-Luke equation. Int. J. Appl. Comput. Math. **11**(2), 35 (2025). https://doi.org/10.1007/s40819-025-01851-8

Adaptive Statistical Round Robin (ASRR): Statistical Measures and Conditional Logic to Optimize System Responsiveness

Isabel Carbonell and Princess Joy Vidad(✉)

College of Information Technology and Computer Science, University of the Cordilleras, Gov. Pack Road, 2600 Baguio City, Benguet, Philippines
princessgvidad@gmail.com

Abstract. Round Robin (RR) is a common CPU scheduling method, but it often becomes slow and unfair when processes have very different burst times. Since RR uses a fixed time quantum, short tasks may wait too long and long tasks may cause too many context switches. This study aims to reduce these problems by creating an improved version of RR called Adaptive Statistical Round Robin (ASRR). The main goal is to adjust the time quantum using simple statistics taken from the current queue of processes. ASRR uses the mean, standard deviation, and minimum burst time to compute a time quantum that better fits the workload. To test the algorithm, 150 simulation batches were created across five different scenarios. ASRR was then compared with Standard RR (StdRR), First-Come First-Serve (FCFS), Min-Max RR (MMRR), and Median-Average RR (MARR). Results show that ASRR gives much faster first responses (ART) when the processes are small and uniform. It also performs well in mixed and random workloads. However, in extreme cases with one very large process, StdRR sometimes performed better because its fixed 25-unit quantum matched the dataset more closely. Overall, the results show that ASRR can reduce starvation and improve responsiveness by adjusting the time quantum based on actual workload behavior.

Keywords: Round Robin Scheduling · Time Quantum · Context Switching · CPU Efficiency · Preemptive Scheduling

1 Introduction

Scheduling algorithms are used by computers to decide which process should run first and how long each process should use the CPU. However, these algorithms often face problems because they need to balance fairness, speed, and efficiency all at once. Moreover, managing trade-offs remains one of the most significant challenges in CPU scheduling.

S. Sodsee et al. (Eds.): IC2IT 2026, LNNS 5590, pp. 35–45, 2026.
https://doi.org/10.1007/978-3-032-28048-0_4

One of these scheduling algorithms is Round Robin (RR), a preemptive algorithm utilizing time quantum. One major problem is that it depends too much on the time quantum, which is the small amount of time given to each process. If the quantum is too small, the CPU wastes a lot of time switching between processes, which leads to additional overhead reducing CPU efficiency [1]. On the other hand, if the time quantum is large, the Average Waiting Time increases and the algorithm starts acting like the First Come First Serve algorithm [2].

Studies show that RR can cause higher waiting and turnaround times, especially when processes have varying burst times. Using a fixed quantum makes the system slower [3]. Compared with enhanced versions, traditional Round Robin performed the worst. In one test with large burst times, it had a very high Average Waiting Time (2319 ms) and Average Turnaround Time of (3023 ms) using a 25 ms time quantum [4].

To solve this, researchers have proposed using a dynamic time quantum (DTQ) that changes depending on the processes' remaining burst times [5,6]. While this helps improve performance, there are still major drawbacks. One significant problem is that many DTQ algorithms depend on the average burst time to decide the time quantum. This becomes unreliable when the processes have very different burst times or include extreme values [2].

This paper aims to design and evaluate an adaptive algorithm that uses descriptive statistics and conditional logic to examine the behavior of the process queue itself [7].

2 Methodology

2.1 Metrics

To comprehensively evaluate the performance of the proposed scheduling algorithm, a standard set of key metrics will be employed. Table 1 lists the literature supporting the selection of these metrics.

Table 1. Key References Supporting Performance Metrics

Metric	Supporting References
AWT	[1,3,4,7–21]
TAT	[1,3,4,7–15,17–21]
ART	[22–33]

Average Waiting Time (AWT) measures the average time processes spend in the ready queue. If the time quantum is large, the Average Waiting Time increases and the algorithm starts acting like the First Come First Serve algorithm, which can make short processes wait too long. It also does not handle priorities well, meaning important tasks might not get done faster than others.

Studies show that RR can cause higher waiting times, especially when processes have varying burst times [2,3].

Average Turnaround Time (TAT) is an important measure of how efficient a scheduling algorithm is. It refers to the total time a process spends in the system, starting from when it is submitted until it is fully completed. Because TAT includes both the time a process waits in the queue and the time it spends using the CPU, it gives a clear picture of how well the scheduler manages tasks. A lower TAT usually means the system is faster and more efficient. Reducing average turnaround time is a main goal in CPU scheduling because it leads to better CPU usage and a more responsive system [1,22].

Average Response Time (ART) is an important metric used to evaluate how interactive and responsive a CPU scheduling algorithm is. It measures the time between when a process arrives in the ready queue and when it first receives CPU time. Since modern operating systems focus heavily on user interaction, minimizing ART is often viewed as more important than reducing total completion time. ART is also very sensitive to the selection of the Time Quantum; choosing an appropriate quantum helps ensure that new processes are not left waiting too long before they begin execution [22,23]. In this simulation, ART was computed programmatically by recording the exact timestamp a process first acquires the CPU. The mathematical definition used is:

$$ART = \frac{1}{n}\sum_{i=1}^{n}(First\_CPU\_Start_i - Arrival\_Time_i) \tag{1}$$

where n represents the total number of processes in the batch. The metric is calculated in milliseconds. For each scenario, the final reported ART is the arithmetic mean of the ARTs across all 30 batches within that scenario.

2.2 Proposed Algorithm (ASRR)

The Adaptive Statistical Round Robin (ASRR) algorithm determines the optimal time quantum based on queue statistics. The detailed logic is described in Algorithm 1, followed by the visual flowchart in Fig. 1.

Algorithm 1. ASRR Time Quantum Calculation

$loadChange \leftarrow (lastN > 0)\ ?\ (|currentN - lastN|/lastN)\ :\ 1.0$
$isStart \leftarrow (index = 0);\ isSpike \leftarrow (loadChange > 0.3);\ hasHist \leftarrow (prevTQ \neq \text{NULL})$
if not $isStart$ **and not** $isSpike$ **and** $hasHist$ **then**
  return $prevTQ$
end if
$B \leftarrow \text{GetBurst}(ReadyQueue)$
$\mu \leftarrow \text{Avg}(B);\ \sigma \leftarrow \text{StdDev}(B);\ minB \leftarrow \text{Min}(B)$
$\tau \leftarrow \mu \times 0.5;\quad k \leftarrow (\mu > 0)?(\sigma/\mu) : 0;\quad \beta \leftarrow minB$
$\alpha \leftarrow 1.0/\sqrt{currentN}$
if $\sigma < \tau$ **then**
  $baseTQ \leftarrow \mu - \sigma$
else
  $baseTQ \leftarrow \mu + (k \times \sigma) + \beta$
end if
$optTQ \leftarrow hasHist\ ?\ ((\alpha \cdot baseTQ) + ((1.0 - \alpha) \cdot prevTQ))\ :\ baseTQ$
return $\lceil \max(optTQ, 1) \rceil$

2.3 Baseline Algorithms

Classic Round Robin (StdRR): The traditional Round Robin scheduling algorithm relies on static quantum rules. The CPU assigns a fixed, static unit of time, called a Time Quantum to every process in the ready queue. The processes are arranged in a First-In-First-Out (FIFO) order. Regardless of the process's priority or the total amount of burst time needed, the CPU chooses the first process in line and executes it strictly for that amount of time.

First-Come First-Serve (FCFS): The traditional FCFS algorithm is the most fundamental non-preemptive scheduling algorithm, using a First-In First-Out basis. Once a process begins execution, it holds it until it terminates with no interruptions. While this approach is simple to implement, and no process is starved endlessly, it is inefficient because of how processes get stuck waiting behind, inflating the average waiting and turnaround times [33].

Min-Max Round Robin (MMRR): The Min-Max Round Robin (MMRR) algorithm begins by sorting the ready queue in ascending order based on burst time [21]. At the start of each scheduling cycle, it calculates the Time Quantum (TQ) by identifying the process with the shortest burst time (Min) and the one with the longest burst time (Max), then computing their arithmetic mean $(TQ = (Max + Min)/2)$.

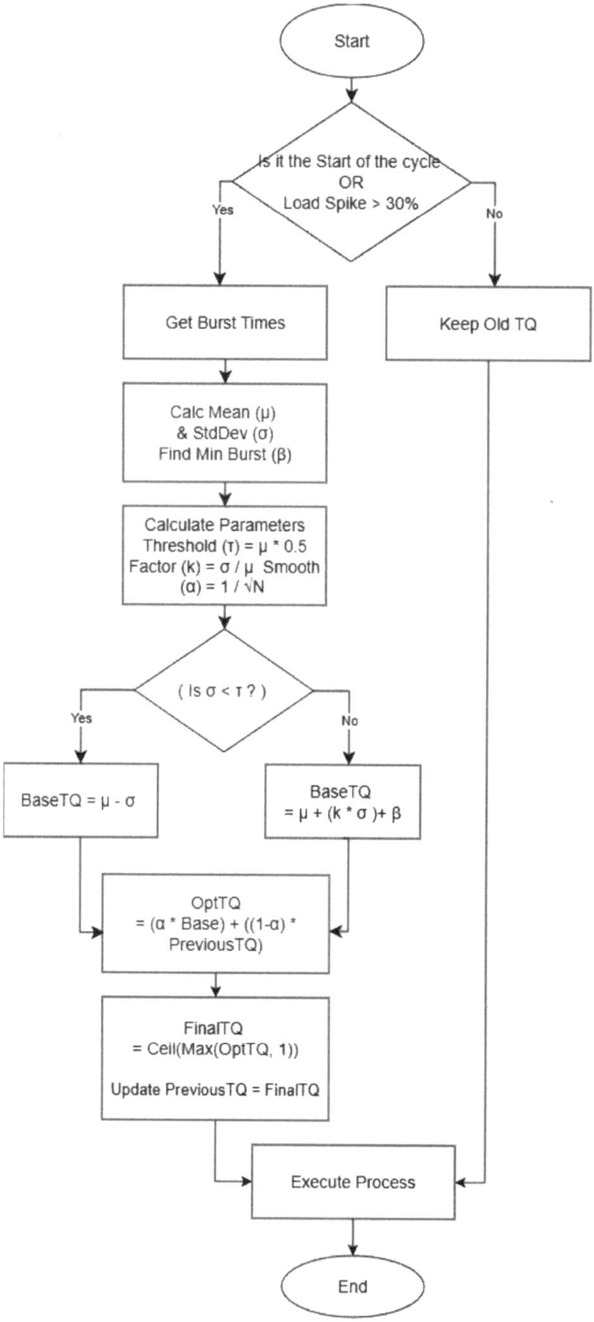

Fig. 1. Adaptive Statistical Round Robin Flowchart

Median-Average Round Robin (MARR): The process begins by organizing the Ready Queue in ascending order. At the start of each cycle, the algorithm calculates a Dynamic Time Quantum (TQ). This is derived by taking the average and the median of the burst times of all available processes and averaging those two values together: $TQ = (Median + Average)/2$.

Comparison: The chosen old algorithms, Round Robin and FCFS, rely on static or mostly static quantum rules. Classic RR uses one constant time slice for all processes, which makes it simple to apply but inefficient when burst times vary widely. The two newer methods (MMRR and MARR) take a more adaptive approach, recalculating the quantum using queue statistics. However, because they depend on averages, these methods become unstable when extreme burst values or sudden arrivals distort the statistics. MMRR, for example, is highly susceptible to outliers; a single process with an extremely large burst time can artificially inflate the quantum, causing the algorithm to degenerate into FCFS behavior [1,6].

3 Findings

To assess the proposed Adaptive Statistical Round Robin (ASRR) algorithm, a controlled simulation was created using Python to measure its performance under different types of workloads. The dataset included 150 simulation batches, which were evenly divided into five different test scenarios, with 30 batches in each scenario. Every batch contained exactly 10 processes. The workloads were designed using specific Burst Time (BT) and Arrival Time (AT) distributions to represent different system conditions. Some scenarios included short processes with burst times between 2 and 5 milliseconds that arrived at the same time to test immediate execution. Other scenarios involved longer processes that required 8 to 15 ms of CPU time and also arrived simultaneously to represent heavier system loads. To examine extreme cases, a convoy effect situation was created by placing one large 45 ms process before nine shorter processes. In addition, randomized workloads were used, where processes had different arrival times and burst times ranging from 2 to 20 ms to imitate unpredictable real-world system activity. To ensure that only the performance of the ASRR algorithm was measured and fairly compared with traditional scheduling methods, the context switching time was set to 0 ms in all scenarios.

3.1 Scenario Analysis

ASRR showed its strongest advantage here, becoming the clear leader in interactivity (ART). ASRR achieved an ART of 12.6 ms, outperforming all others. This was due to the low standard deviation in the uniform workload, which allowed ASRR to produce a highly optimized Time Quantum (Fig. 2).

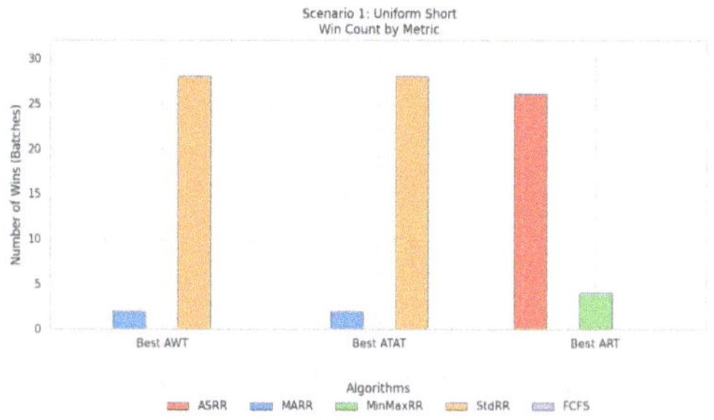

Fig. 2. Scenario 1 Algorithm Comparison (ART Dominance)

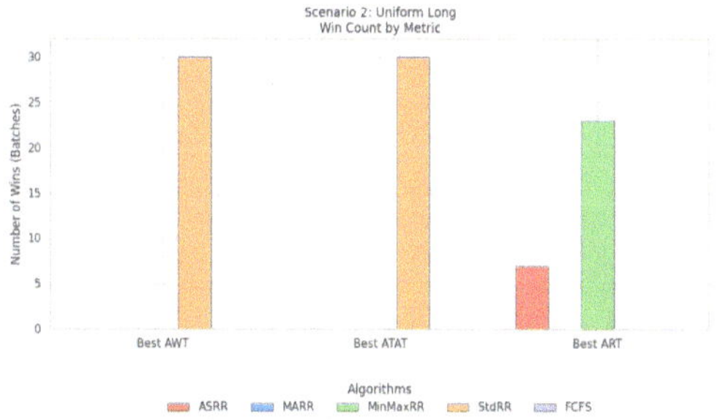

Fig. 3. Scenario 2 Algorithm Comparison (Heavy Load)

Under heavy load conditions with long and uniform burst times, Standard Round Robin (StdRR) dominated the efficiency metrics. Min-Max RR performed the best in Response Time. ASRR's smoothing method produced a Time Quantum that was slightly larger than ideal here, trading initial speed for stability (Fig. 3).

3.1.3 Scenario 3: Long AT and Short BT (Outliers) This scenario included one very large job and many extremely small ones ("Convoy Effect"). StdRR performed best because its fixed 25 ms quantum happened to match the dataset needs. ASRR detected high variance and increased the quantum to reduce context switching, which inadvertently delayed the smaller jobs behind the large one (Fig. 4).

3.1.4 Scenario 4: Long AT and Long BT (Mixed High Traffic) MARR performed best in ART with ASRR following closely. This outcome is important because it shows that in high-traffic environments with a wide range of process sizes, ASRR's statistical method helps it avoid the "convoy effect" more effectively than many traditional approaches (Fig. 5).

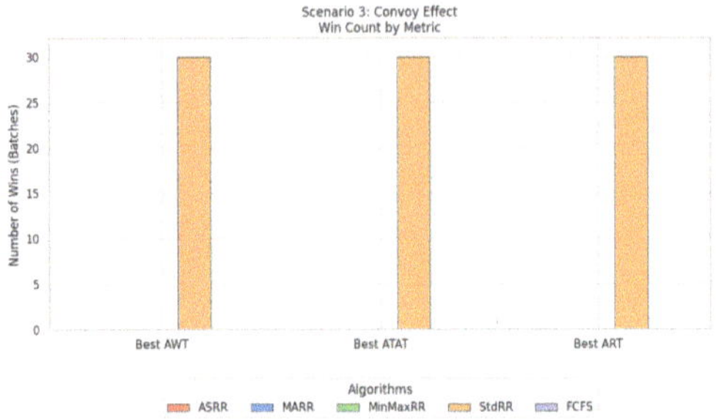

Fig. 4. Scenario 3 Algorithm Comparison (Outlier Impact)

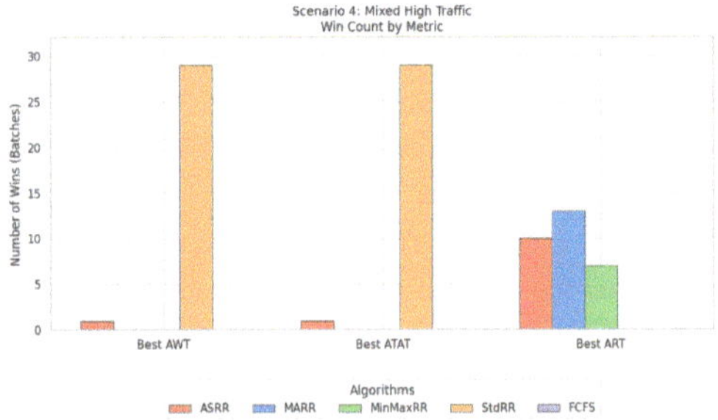

Fig. 5. Scenario 4 Algorithm Comparison (Mixed Traffic)

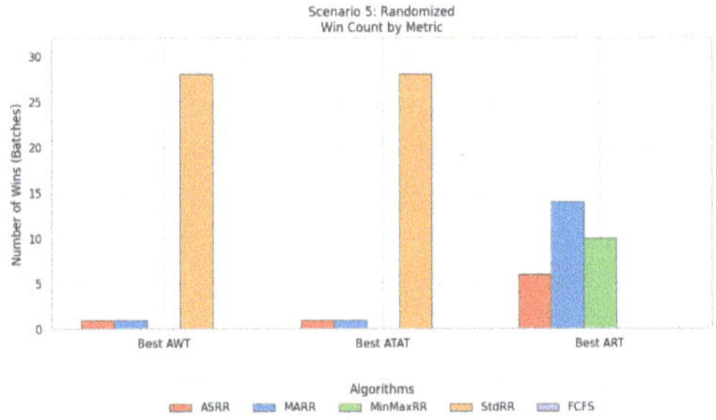

Fig. 6. Scenario 5 Algorithm Comparison (Randomized Workload)

3.1.5 Scenario 5: Randomized AT and BT In unpredictable workloads, StdRR dominated Waiting Time, but the competition for Response Time (ART) was intense. ASRR consistently remained competitive, unlike FCFS and StdRR which recorded significantly worse interactivity in some batches (Fig. 6 and Table 2).

Table 2. Summary of Average Results across all Scenarios (Time in ms)

Scenario	Metric	FCFS	StdRR	MMRR	MARR	ASRR
1. Short/Short (Uniform)	TAT	20.4	20.4	25.2	21.9	27.3
	AWT	16.6	16.6	21.4	18.1	23.5
	ART	16.6	16.6	14.1	16.0	**12.6**
2. Short/Long (Heavy Load)	TAT	57.9	85.6	90.4	76.6	81.7
	AWT	47.7	75.4	80.2	66.4	71.5
	ART	47.7	**22.5**	33.1	39.4	39.2
3. Long/Short (Outliers)	TAT	58.5	**25.3**	63.4	60.8	58.5
	AWT	51.2	18.0	56.1	53.5	51.2
	ART	51.2	15.2	48.7	50.0	51.2
4. Long/Long (Mixed Traffic)	TAT	51.9	65.4	54.4	53.5	62.0
	AWT	42.2	55.7	44.7	43.8	52.3
	ART	42.2	19.9	37.9	36.9	38.5
5. Random (Unpredictable)	TAT	61.1	75.1	69.8	68.8	67.2
	AWT	49.7	63.7	58.4	57.4	55.8
	ART	49.7	**20.1**	44.9	43.7	48.9

4 Conclusion

The Adaptive Statistical Round Robin (ASRR) algorithm was developed to mitigate long process delays. Unlike modern schedulers such as the Earliest Eligible Virtual Deadline First (EEVDF) in the Linux kernel or Shortest Remaining Time First (SRTF), ASRR maintains the inherent simplicity of traditional Round Robin while introducing statistical workload awareness.

The benefits of this variance-based quantum scaling are quantitatively evident. In workloads with uniform and short processes, ASRR achieved an Average Response Time (ART) of 12.6 milliseconds, a **24.1% improvement** over Standard Round Robin (StdRR). Additionally, in unpredictable, randomized workloads, ASRR reduced Average Waiting Time (AWT) by **12.4%** and Average Turnaround Time (TAT) by **10.5%**, proving its efficacy in dynamic environments.

While ASRR successfully reduces starvation-like delays, this adaptability introduces a computational trade-off. Unlike StdRR's static $\mathcal{O}(1)$ scheduling overhead, ASRR requires an $\mathcal{O}(n)$ time complexity per cycle to recalculate the mean and standard deviation of the ready queue. Although minimal for typical modern queue sizes, this $\mathcal{O}(n)$ recalculation must be considered in highly saturated systems. Overall, ASRR demonstrates that basic statistical methods and conditional logic can effectively optimize CPU scheduling in imbalanced computing environments.

References

1. Qazi, F., et al.: Improving round robin scheduling with dynamic time quantum (IRRDQ). J. Appl. Eng. Technol. (JAET) **7**(2), 70–82 (2023)
2. Olofintuyi, S.S., et al.: A survey of variants of round robin CPU scheduling algorithms. FUDMA J. Sci. **4**(4), 526–546 (2021)
3. Zohora, M.F., et al.: An enhanced round robin using dynamic time quantum for real-time asymmetric burst length processes. PLoS ONE **19**(8), e0304517 (2024)
4. Fiad, A., et al.: Improved version of round robin scheduling algorithm based on analytic model. Int. J. Netw. Distrib. Comput. **8**, 195–202 (2020)
5. Najafi, S., Noferesti, S.: Determining dynamic time quantum in round-robin scheduling algorithm using machine learning. Soft Comput. J. **10**(2), 32–43 (2022)
6. Iqbal, M., et al.: Optimizing task execution: the impact of dynamic time quantum and priorities on Round Robin scheduling. Future Internet **15**(3), 104 (2023)
7. Al Refai Mohammed, N., Al-Qawasmi, K., Abdulkarim, D.A., Haithem, A.W., Otoom, A., Al-Momani, A.D.A.I.: Investigating the impact of statistical measures on round robin quantum length selection in scheduling algorithms. WSEAS Trans. Comput. Res. **13**, 450–461 (2025)
8. Putra, T.D., Purnomo, R.: Simulation of priority round robin scheduling algorithm. SinkrOn **7**(4), 2170–2181 (2022)
9. Putra, T.D., Purnomo, R.: Average max round robin algorithm: a case study. SinkrOn **8**(4), 2662–2669 (2023)
10. Mirkar, S.Q., Bobal, A., Gattani, S.: Agile time quantum round robin algorithm. In: IEEE, pp. 1–7 (2024)
11. Putra, M.T.D., Hidayat, H., Septian, N., Afriani, T.: Analisis perbandingan Algoritma penjadwalan CPU first come first serve (FCFS) dan round robin. Build. Inf. Technol. Sci. (BITS) **3**(3), 207–212 (2021)
12. Sahoo, B.K.: Novel approach of diagnosing significant metrics of load balancing using CloudSim. In: 2022 10th International Conference on Emerging Trends in Engineering and Technology, pp. 1–6. IEEE (2022)
13. Abdelkader, A.: VORR: a new round robin scheduling algorithm. Al-Azhar Bull. Sci. (2021)
14. Sakshi, N., et al.: A new median-average round Robin scheduling algorithm: an optimal approach for reducing turnaround and waiting time. Alex. Eng. J. **61**(12), 10527–10538 (2022)
15. Putra, T.D., Purnomo, R.: Case study: improved round robin algorithm. SinkrOn **7**(3), 950–956 (2022)
16. Purnomo, R., Putra, T.D.: Comparison between simple round robin and improved round robin algorithms. JATISI **9**(3), 2205–2221 (2022)
17. Moni, M.M.A.: Comparative analysis of process scheduling algorithm using AI models. In: ICCIT, pp. 587–592. IEEE (2022)
18. Abubakar, S.E.: Modified round robin with highest response ratio next CPU scheduling algorithm using dynamic time quantum. SLU J. Sci. Technol. 87–99 (2023)
19. Gupta, A. K., et al.: ORR. In: International Conference on Data Science, Machine Learning and Artificial Intelligence, pp. 296–304 (2021)
20. Mangukia, A.: Improved variable round robin scheduling algorithm. In: International Conference on Computing Communication and Networking Technologies, pp. 1–7 (2021)

21. Panda, S.K., Bhoi, S.K.: An effective round robin algorithm using min-max dispersion measure. Int. J. Comput. Appl. **51**(1), 45–49 (2012)
22. Bandara, A.M.O., Rajapaksha, U.U.S.: An improved round robin algorithm for an efficient CPU scheduling. In: ICARC, pp. 349–354 (2022)
23. Abdelhamid, A.M.H., Zamel, A.A.: Implementing and measuring the performance of PB, RR and PBRR scheduling algorithms. In: Novel Intelligent and Leading Emerging Sciences Conference, pp. 18–22 (2023)
24. Tang, Y.: Response time analysis of lazy round robin. Semant. Scholar 258–263 (2021)
25. Yuan, Y., et al.: Adaptive priority adjustment scheduling approach with response-time analysis in time-sensitive networks. IEEE Trans. Ind. Inf. **18**(12), 8714–8723 (2022)
26. Yuan, H., et al.: Online dispatching and fair scheduling of edge computing tasks: a learning-based approach. IEEE Internet Things J. **8**(19), 14985–14998 (2021)
27. Arafat, A.A.: Response time analysis for dynamic priority scheduling in ROS2. In: Proceedings of the 59th ACM/IEEE Design Automation Conference, pp. 301–306 (2022)
28. Tang, Y.: Response time analysis and priority assignment of processing chains on ROS2 executors, pp. 231–243. IEEE (2020)
29. Wang, Z., et al.: Deep reinforcement learning-based scheduling for optimizing system load and response time in edge and fog computing environments. Futur. Gener. Comput. Syst. **152**, 55–69 (2023)
30. Blaß, T.: A ROS 2 response-time analysis exploiting starvation freedom and execution-time variance, pp. 41–53. IEEE (2021)
31. Khajeh, S.A., et al.: Real-time scheduling in IoT applications: a systematic review. Sensors **23**(1), 232 (2022)
32. Jiang, X.: Real-time scheduling and analysis of processing chains on multi-threaded executor in ROS 2, pp. 27–39. IEEE (2022)
33. Agcaoili, A.: CPU scheduling: a study of the first come first serve and round robin approaches. ResearchGate (2022)

A Lightweight Transformer–Enhanced Image Processing Framework for Efficient Visual Feature Representation

Van Van Chau[1], Tuan Minh Nguyen[1], and Phayung Meesad[2](✉)

[1] Faculty of Information Technology, Posts and Telecommunications Institute of Technology, 11 Nguyen Dinh Chieu, Ho Chi Minh City 700000, Vietnam
{vancv,minhtuan}@ptit.edu.vn

[2] Department of Information Technology and Management, King Mongkut's University of Technology North Bangkok, Bangsue 10800, Thailand
phayung.m@itd.kmutnb.ac.th

Abstract. Transformer-based models have become increasingly prominent in image processing due to their ability to capture long-range spatial relationships [1]. However, their self-attention mechanism scales quadratically with the number of tokens, resulting in increased memory usage and inference latency that limit practical deployment in real-time or resource-constrained settings [2]. This study investigates whether transformer efficiency can be improved without sacrificing representational capability. We propose a lightweight transformer-enhanced framework that combines a reduced-depth Vision Transformer backbone with a compact feature refinement module. The proposed architecture aims to preserve spatial interaction modeling while operating under a controlled parameter budget. Experiments conducted on CIFAR-10 and Corel-1K demonstrate that the proposed model achieves higher accuracy and F1-score than a conventional CNN baseline, while requiring substantially lower inference time than a full-scale ViT. The results confirm that structural moderation and targeted refinement can maintain effective visual representation learning with improved computational practicality.

Keywords: ViTs · CNNs · Deep Learning · Image Processing · Lightweight Transformer

1 Introduction

Image classification systems have evolved substantially with the maturation of deep neural architectures. Convolutional Neural Networks (CNNs) remain widely adopted due to their efficient hierarchical feature extraction and strong empirical reliability across visual benchmarks [3]. Their localized receptive fields and parameter sharing contribute to stable optimization. However, spatial reasoning in CNNs is inherently constrained by kernel locality. Extending receptive fields through deeper stacking increases model depth and computational load,

S. Sodsee et al. (Eds.): IC2IT 2026, LNNS 5590, pp. 46–55, 2026.
https://doi.org/10.1007/978-3-032-28048-0_5

yet does not always translate into more effective long-range interaction modeling. Transformer-based architectures offer a different formulation by treating images as sequences of tokens. Through self-attention, spatial positions can interact directly, enabling flexible dependency modeling across the image domain. Despite these advantages, the quadratic scaling of attention with respect to token number introduces significant computational cost. As discussed in [4], full-scale Vision Transformer (ViT) models often incur increased memory usage and slower inference, limiting suitability for latency-sensitive or resource-constrained environments. Addressing this constraint has become a central design consideration. Prior efforts include architectural simplification, modified attention schemes, and hybrid CNN–transformer integrations aimed at moderating runtime demand while retaining expressive capacity [5]. In parallel, lightweight feature recalibration modules have been proposed to selectively emphasize informative activations with modest additional computation [6]. Nevertheless, backbone reduction and feature refinement are frequently investigated in isolation. Their combined impact on accuracy-latency trade-offs has received comparatively limited attention. In this work, we explore a compact transformer-enhanced framework that integrates a reduced-depth ViT backbone with a structured refinement component [7]. Rather than maximizing parameter count, the design emphasizes controlled capacity and selective feature strengthening. Empirical evaluation on benchmark datasets indicates that the proposed configuration improves predictive performance relative to a CNN baseline while maintaining lower inference latency than a conventional full-scale ViT under comparable conditions. These observations motivate a closer examination of how moderated transformer depth and lightweight recalibration interact to support efficient visual representation learning. The remainder of this paper is organized as follows. Section 2 reviews related research on efficient transformer architectures and hybrid designs. Section 3 presents the proposed methodology. Section 4 details experimental settings and quantitative results. Section 5 concludes the study and discusses future directions.

2 Related Work

The use of token-based self-attention in visual recognition demonstrated that convolution is not the only structural choice for large-scale modeling [1]. By representing images as sequences, transformer architectures allow interactions across spatial positions without relying solely on progressively enlarged receptive fields. Subsequent investigations focused on how such architectures behave under practical constraints, particularly when scaling to higher resolutions or limited computational budgets [3,8]. One immediate challenge is the quadratic growth of attention computation as token numbers increase. In realistic deployment scenarios, this scaling characteristic translates directly into higher memory requirements and slower inference. As a result, multiple strategies have emerged to reduce structural redundancy or restrict interaction scope [4]. Window-based formulations, including the Swin Transformer, confine attention to localized regions and shift the windows between layers, thereby improving scalability while maintaining competitive recognition results [9]. Compact configurations introduce additional considerations. Knowledge distillation has been applied to stabilize smaller transformer variants and improve convergence behavior. Meanwhile, scaling studies examine how architectural depth and embedding dimensions influence the marginal gains obtained from added complexity [10]. In several cases, increasing capacity yields diminishing returns when runtime constraints are imposed. Another direction addresses representation refinement rather than backbone restructuring. Channel-and spatial-recalibration mechanisms such as CBAM and Squeeze-and-Excitation Networks adjust feature responses dynamically to emphasize informative activations [6,7]. When combined with reduced transformer depth, such modules can partially offset representational loss without significantly expanding parameter count. Integration of convolutional filtering and attention-based interaction has also been explored. CNN-ViT hybrids indicate that localized feature extraction and token-level aggregation may complement each other under constrained resource settings [11,12]. Design choices derived from mobile CNN research further inform transformer adaptation for deployment-aware environments [13]. Beyond architectural efficiency, robustness considerations have gained attention. Empirical studies in domains including medical imaging, noisy-label learning, and alternative attention formulations highlight that runtime reduction alone does not guarantee stable performance [14–16]. More recent mobile-oriented transformer variants reorganize attention computation to better accommodate edge inference conditions [17–19].

3 Our Proposed Method

The proposed model is structured around a practical constraint: increasing transformer depth does not necessarily yield proportional gains once inference cost becomes a limiting factor. Rather than expanding attention layers, we adopt a reduced-capacity configuration and introduce a lightweight refinement component to compensate for representational loss. The overall workflow consists of four stages: input normalization, compact transformer encoding, feature recalibration, and classification. Input images are first resized to a fixed spatial resolution to ensure consistent feature statistics during training. This preprocessing step avoids scale-induced instability and facilitates batch optimization. A shallow transformer encoder then processes the tokenized representation. By limiting the number of attention blocks, the model reduces computational demand while still enabling interaction across spatial regions. This configuration prioritizes controlled contextual aggregation instead of architectural depth expansion. Following encoding, a refinement stage adjusts channel responses through adaptive weighting. Not all feature dimensions contribute equally to prediction; therefore, recalibration is applied to amplify informative activations and suppress weaker responses. The refinement module operates with minimal additional parameters, serving as a corrective mechanism for the reduced backbone capacity. The resulting representation is subsequently passed to the classifier. As illustrated in Fig. 1, the architecture balances contextual interaction and computational moderation without relying on excessive layer stacking. The design reflects an intentional trade-off: constrained depth paired with selective enhancement to maintain discriminative capability under deployment-aware conditions (Table 1).

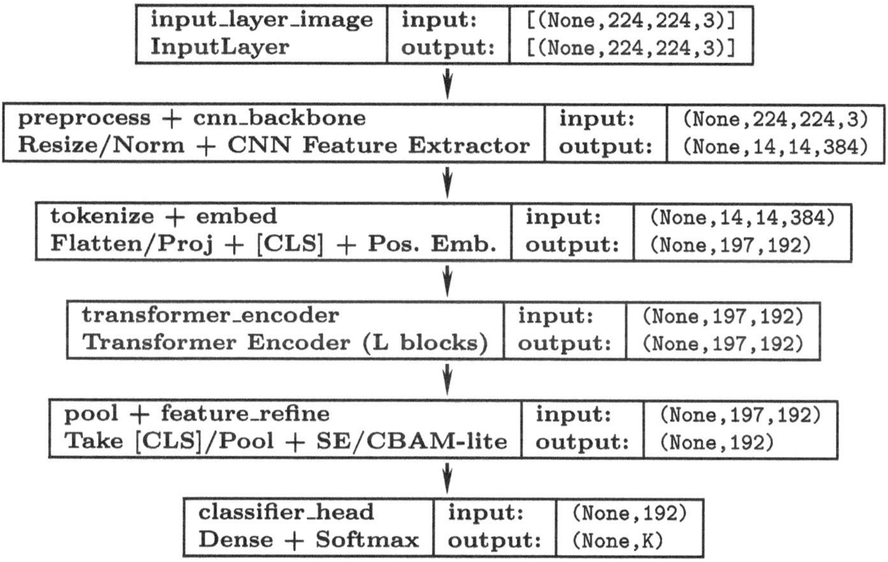

Fig. 1. Model Architecture.

Table 1. Model settings.

Model	Input	Output	Layers	Params
CNN Baseline	(224,224,3)	(K)	Conv + FC (8 layers)	4.3M
Full ViT	(224,224,3)	(K)	12 Transformer Blocks	86M
Proposed CNN–ViT	(224,224,3)	(K)	CNN + L Transformer + Refinement	18.7M

3.1 Image Preprocessing

let the input image be denoted as $I \in \mathbb{R}^{H \times W \times C}$. Before feature extraction, several preparatory steps are applied to regulate optimization behavior. All images are resized to a predefined spatial dimension $H' \times W'$, so that inputs share a uniform scale. Pixel values are normalized according to the statistical distribution of the training set, which helps maintain well-conditioned updates during optimization. During training, data augmentation may additionally be introduced to broaden variability within the samples, and these transformations reduce the influence of lighting fluctuations and contrast variation, allowing the network to concentrate on structural patterns rather than low-level intensity inconsistencies.

3.2 Lightweight Transformer-Based Feature Extraction

The standardized image is decomposed into N non-overlapping patches of spatial dimension $P \times P$. Each patch is transformed into a vector and mapped to a D-dimensional latent space through a learnable projection, resulting in an ordered embedding sequence. To summarize image-level information within the same sequence, a dedicated classification token is prepended, and positional embeddings are injected to retain spatial ordering. The resulting tokenized representation is written as

$$Z_0 = [z_{\text{cls}}; z_1; z_2; \ldots; z_N] + E_{\text{pos}}, \tag{1}$$

where E denotes the patch projection matrix, E_{pos} represents positional embeddings, and N is the number of image patches.

This sequence is then propagated through L compact transformer blocks. Each block alternates multi-head self-attention for contextual interaction and a feed-forward subnetwork for feature transformation, both coupled with residual connections:

$$Z'_\ell = \text{MSA}(\text{LN}(Z_{\ell-1})) + Z_{\ell-1}, \tag{2}$$

$$Z_\ell = \text{FFN}(\text{LN}(Z'_\ell)) + Z'_\ell, \tag{3}$$

where $\text{LN}(\cdot)$ denotes layer normalization. In contrast to full-scale ViTs designs, we reduce the depth and attention capacity to control compute while retaining effective global dependency modeling [16].

Algorithm 1. Feature Extraction and Refinement using Lightweight ViTs

Input: I // a set of input images
Output: y refined global feature vectors (classification token)
Resize image I to 224×224
Split I into N non-overlapping patches $\{p_1, \ldots, p_N\}$ of size $P \times P \times C$
for $i = 1$ to N **do**
  Flatten patch $p_i \in \mathbb{R}^{P \times P \times C}$ to vector $x_i \in \mathbb{R}^{P^2 \cdot C}$
  Project: $z_i \leftarrow W_E x_i + b_E$
end for
Concatenate tokens: $Z \leftarrow [z_{\text{cls}}; z_1; \ldots; z_N]$
Add positional encoding: $Z \leftarrow Z + E_{\text{pos}}$
for $\ell = 1$ to L **do**
  $Z \leftarrow \text{MSA}(\text{LN}(Z)) + Z$
  $Z \leftarrow \text{FFN}(\text{LN}(Z)) + Z$
end for
Extract global feature: $f \leftarrow \text{LN}(z_{\text{cls}})$
Compute refinement weights: $w \leftarrow \sigma(W_r f + b_r)$
Refine feature: $\hat{f} \leftarrow w \odot f$
return $\hat{f}$

Most computational overhead arises from the transformer encoder, as processing complexity increases with patch number and attention heads. The reduced-depth configuration and added refinement component shorten inference time compared to larger transformer architectures, while sustaining effective visual feature encoding.

3.3 Feature Refinement Module

Global contextual modeling from the transformer encoder produces high-dimensional representations, yet their contributions to prediction are not uniformly distributed. A lightweight refinement mechanism is applied to adjust feature responses in a controlled manner. Channel-wise weighting is applied prior to normalization to adaptively adjust feature activations. For a feature vector f, the refinement process produces an updated representation $\hat{f}$ expressed by

$$\hat{f} = \sigma(W_r f + b_r) \odot f, \tag{4}$$

The parameters W_r and b_r are optimized during training. The sigmoid function $\sigma(\cdot)$ produces channel-wise scaling factors, and $\odot$ applies element-wise multiplication. This adjustment emphasizes channels aligned with the target objective while suppressing weaker responses. The refinement step incurs modest computational expense and contributes to more discriminative feature encoding.

3.4 Training Objective

Model parameters are learned under a standard supervised setting. Given K target classes, optimization minimizes the cross-entropy loss

$$\mathcal{L}_{\text{cls}} = -\sum_{i=1}^{K} y_i \log(\hat{y}_i). \tag{5}$$

Here, y_i represents the ground-truth label and $\hat{y}_i$ denotes the predicted probability for class i. The transformer encoder and the refinement module are updated jointly, enabling their feature representations to co-adapt throughout training rather than being tuned in isolation.

Optimization proceeds using mini-batch gradient descent. Instead of relying solely on final accuracy, convergence behavior is monitored during training to avoid unstable updates. After training stabilizes, performance is assessed on a held-out validation split. In addition to classification accuracy, average inference time is recorded to reflect runtime characteristics. Detailed optimization and evaluation steps are summarized in Algorithm 2.

Algorithm 2. Joint Optimization and Evaluation Procedure

Input: Training set $\mathcal{D}_{tr}$, validation set $\mathcal{D}_{val}$
Output: Optimized parameters θ^*, Accuracy, F1-score, Latency

```
Setting parameters θ
repeat for each epoch
    Draw a mini-batch (I, y) from D_tr
    Compute feature representations y_i ← Φ_θ(I) via Algorithm 1
    Obtain predictions ŷ ← Softmax(W y_i + b)
    Adjust θ by backpropagating the cross-entropy objective
until convergence criterion is met
Assess model performance on D_val to obtain Accuracy and F1-score
Record mean inference latency (ms/image) on D_val
return θ*, Accuracy, F1-score, Latency
```

4 Experiments and Results

This section evaluates the proposed lightweight transformer–enhanced framework on image classification tasks. Experiments are conducted on CIFAR-10[1] and Corel-1K[2], two widely used benchmark datasets, under a consistent and reproducible experimental setup. All experiments were conducted on a single GPU environment to ensure consistent latency measurement. Dataset characteristics and train–test splits are reported in Table 2. Images are resized and normalized prior to training, with standard data augmentation applied. All models are trained using the Adam optimizer with a cross-entropy loss. Performance is assessed using accuracy, F1-score, and average inference latency.

Table 2. Datasets used in the experiments.

Dataset	#Classes	#Images	Image Size	Split (Train/Test)
CIFAR-10	10	60,000	32×32	50k/10k
Corel-1K	10	1,000	Variable	800/200

For both datasets, images are resized and normalized prior to training, with standard data augmentation applied. All models are trained using a cross-entropy loss optimized with Adam, and evaluated in terms of accuracy, F1-score, and average inference latency. All comparison models were implemented using an identical training and evaluation setup to ensure a fair assessment. performance results are summarized in Table 3. For both CIFA-10 and Corel-1K, our

[1] A. Krizhevsky and G. Hinton, *Learning Multiple Layers of Features from Tiny Images*, Technical Report, University of Toronto, 2009.

[2] J. Z. Wang, J. Li, and G. Wiederhold, *SIMPLIcity: Semantics-sensitive Integrated Matching for Picture Libraries*, IEEE Transactions on Pattern Analysis and Machine Intelligence, 2001.

work achieves higher accuracy and F1-score than the CNN baseline. In addition, its inference time is substantially shorter than that of the full-scale ViT. Its observations indicate that the model enhances predictive capability while keeping processing demands at a controlled level.

Table 3. Performance comparison accross models based on datasets.

Dataset	Method	Accuracy (%)	F1-score (%)	Latency (ms)
CIFAR-10	CNN Baseline [20]	74.2	73.5	41.8
	ViTs baseline [5]	88.6	87.9	92.4
	Our Model	**91.4**	**90.8**	**38.1**
Corel-1K	CNN Baseline [20]	70.8	69.9	39.6
	ViTs baseline [5]	83.7	82.5	86.3
	Our Model	**87.1**	**86.2**	**33.9**

5 Discussion and Evaluation

Figure 2 illustrates the training trajectory of the proposed model. A rapid reduction in loss is observed during the initial epochs, followed by a more gradual decrease as optimization progresses. This transition suggests that the model quickly captures dominant patterns before entering a refinement phase. The validation curve exhibits a consistent upward trend with limited oscillation, indicating stable convergence rather than erratic updates. No sudden spikes or divergence are observed, which supports the reliability of the adopted capacity configuration.

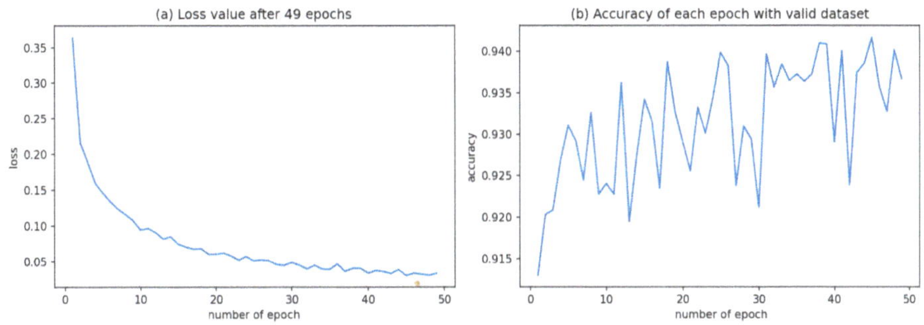

Fig. 2. Our experimental results.

When compared with baseline configurations, the proposed framework achieves improved accuracy and F1-score relative to CNN counterparts while maintaining shorter inference time than a full-scale ViT under identical conditions. This outcome reflects a deliberate structural compromise: transformer depth is moderated to contain computational cost, and the refinement stage compensates by enhancing discriminative channel responses. Although larger ViT models can yield marginally higher peak accuracy, their runtime overhead increases noticeably, which constrains practical deployment in latency-sensitive environments.

6 Conclusion and Future Work

The presented study investigates whether transformer depth can be moderated without forfeiting discriminative capability in image classification. Empirical evaluation indicates that a restrained attention configuration, when coupled with targeted feature recalibration, sustains competitive predictive performance under reduced computational demand. In direct comparison settings, the proposed framework exceeds the CNN baseline in accuracy and F1-score and exhibits shorter inference time than a conventional full-scale ViT. These observations suggest that contextual interaction does not depend solely on architectural expansion. Instead, selective enhancement of informative feature channels appears sufficient to preserve representational strength within a compact structure. The recorded training dynamics further reflect stable convergence behavior rather than oscillatory updates. Several aspects remain open for further investigation. Evaluation across larger and more heterogeneous datasets is required to verify generalization beyond the current benchmarks. Additionally, refinement mechanisms can be re-examined to determine whether alternative lightweight modulation strategies offer improved trade-offs. Finally, deployment-oriented validation on edge hardware platforms will clarify operational behavior under strict runtime constraints.

References

1. Dosovitskiy, A., et al.: An image is worth 16×16 words: transformers for image recognition at scale. arXiv, (2020). abs/2010.11929
2. Papa, L., Russo, P., Amerini, I., Zhou, L.: A survey on efficient visiontransformers: algorithms, techniques, and performance benchmarking. IEEE Trans. Pattern Anal. Mach. Intell. **46**(12), 7682–7700 (2024). https://doi.org/10.1109/TPAMI.2024.3392941
3. Khan, S., Naseer, M., Hayat, M., Zamir, S.W., Khan, F.S., Shah, M.: Transformers in vision: a survey. ACM Comput. Surv. **54**(10s), 1–41 (2022). https://doi.org/10.1145/3505244
4. Saha, S., Xu, L.: Vision transformers on the edge: a comprehensive survey of model compression and acceleration strategies. Neurocomputing **643**, 130417 (2025). https://doi.org/10.1016/j.neucom.2025.130417
5. Chen, X., Cao, Q., Zhong, Y., Zhang, J., Gao, S., Tao, D. : DearKD: data-efficient early knowledge distillation for vision transformers. In: 2022 IEEE/CVF Conference on Computer Vision and Pattern Recognition (CVPR), pp. 12042–12052.

IEEE, New Orleans, LA, USA (2022). https://doi.org/10.1109/CVPR52688.2022.01174
6. Woo, S., Park, J., Lee, J.-Y., Kweon, I.S.: CBAM: convolutional block attention module. In: Ferrari, V., Hebert, M., Sminchisescu, C., Weiss, Y. (eds.) ECCV 2018. LNCS, vol. 11211, pp. 3–19. Springer, Cham (2018). https://doi.org/10.1007/978-3-030-01234-2_1
7. Hu, J., Shen, L., Albanie, S., Sun, G., Wu, E.: Squeeze-and-excitation networks. IEEE Trans. Pattern Anal. Mach. Intell. **42**(8), 2011–2023 (2020). https://doi.org/10.1109/TPAMI.2019.2913372
8. Lin, T., Wang, Y., Liu, X., Qiu, X.: A survey of transformers. AI Open **3**, 111–132 (2022). https://doi.org/10.1016/j.aiopen.2022.10.001
9. Liu, Z., et al.: Swin transformer: hierarchical vision transformer using shifted windows. In: 2021 IEEE/CVF International Conference on Computer Vision (ICCV), pp. 9992–10002. IEEE, Montreal, QC, Canada (2021). https://doi.org/10.1109/ICCV48922.2021.00986
10. Zhai, X., Kolesnikov, A., Houlsby, N., Beyer, L.: Scaling vision transformers. In: 2022 IEEE/CVF Conference on Computer Vision and Pattern Recognition (CVPR), pp. 1204–1213. IEEE, New Orleans, LA, USA (2022). https://doi.org/10.1109/CVPR52688.2022.01179
11. Cheng, S., Xiao, X., Wang, L.: Deep hashing image retrieval based on CNN and visual transformer network. Appl. Soft Comput. **177**, 113244 (2025). https://doi.org/10.1016/j.asoc.2025.113244
12. Zhang, Y., Wei, L., Zhang, B., Liu, Z., Yi, K., Hu, S.: A lightweight convolution and vision transformer integrated model with multi-scale self-attention mechanism. Neurocomputing, **658** (2025). https://doi.org/10.1016/j.neucom.2025.131670
13. Howard, A., et al.: Searching for MobileNetV3. In: 2019 IEEE/CVF International Conference on Computer Vision (ICCV), pp. 1314–1324. IEEE, Seoul, Korea (South) (2019). https://doi.org/10.1109/ICCV.2019.00140
14. Al-hammuri, K., Gebali, F., Kanan, A., Chelvan, I.T.: Vision transformer architecture and applications in digital health: a tutorial and survey. Vis. Comput. Ind. Biomed. Art **6**(1), 14 (2023). https://doi.org/10.1186/s42492-023-00140-9
15. Mots'oehli, M., Mogale, H., Baek, K.: Balancing accuracy, calibration, and efficiency in active learning with vision transformers under label noise. CoRR, (2025). abs/2505.04375. https://doi.org/10.48550/ARXIV.2505.04375
16. Kim, H., Ko, B.C.: Rethinking attention mechanisms in vision transformers with graph structures. Sensors **24**(4), 1111 (2024). https://doi.org/10.3390/s24041111
17. Mehta, S., Rastegari, M.: MobileViT: light-weight, general-purpose, and mobile-friendly vision transformer. arXiv (2021). abs/2110.02178
18. Graham, B., et al.: LeViT: a vision transformer in ConvNet's clothing for faster inference. In: 2021 IEEE/CVF International Conference on Computer Vision (ICCV), pp. 12239–12249. IEEE, Montreal, QC, Canada (2021). https://doi.org/10.1109/ICCV48922.2021.01204
19. Truong-Dang, V.-L., Thai, H.-T., Le, K.-H.: TinyResViT: a lightweight hybrid deep learning model for on-device corn leaf disease detection. Internet Things **30**, 101495 (2025). https://doi.org/10.1016/j.iot.2025.101495
20. Kolhe, P.B., Shelke Ramesh, D., Agarwal, N.: EnhanceNet: rethinking model scaling for convolutional neural network. In: Iglesias, A., Shin, J., Patel, B., Joshi, A., (eds.) Information Systems for Intelligent Systems, pp. 1–15. Springer, Cham (2025)

Classification of Pet-Friendly Hotel Reviews on Online Platforms Using Text Mining Techniques

Supassara Heebthamai(✉) and Watchareewan Jitsakul

Faculty of Information Technology and Digital Innovation, King Mongkut's University of Technology North Bangkok, Bangkok, Thailand
s6707021856132@email.kmutnb.ac.th,
watchareewan.j@itd.kmutnb.ac.th

Abstract. This research aims to classify user opinions of pet-friendly hotel services using a dataset of 1,000 customer reviews collected from the Agoda platform. The reviews were categorized into positive and negative sentiments. The data were processed through text cleaning, Thai word segmentation, and numerical vectorization. Subsequently, the dataset was divided and evaluated using 5-fold cross-validation to compare the performance of deep learning techniques based on natural language processing models, including Recurrent Neural Networks (RNN), Bidirectional Recurrent Neural Networks (Bi-RNN), Long Short-Term Memory networks (LSTM), and Bidirectional Long Short-Term Memory networks (Bi-LSTM). The experimental results indicate that the Bidirectional Long Short-Term Memory (Bi-LSTM) model achieved the highest performance. When evaluated on unseen data, the model achieved 86.67% accuracy, 0.87 precision, 0.87 recall, and 0.87 F1-score.

Keywords: Hotel Review · Text Mining · Bi-LSTM · LSTM · Bi-RNN · RNN

1 Introduction

In recent years, traveling with pets has gained increasing popularity, particularly among consumers who regard their pets as family members. This trend has led to a growing demand for accommodations that allow pets, resulting in a significant expansion of pet-friendly hotels in both major cities and popular tourist destinations across Thailand. As a consequence, pet-friendly hotels have emerged as an important segment within the hospitality industry, offering new business opportunities and competitive advantages.

Customers frequently share their experiences through online review platforms, providing valuable insights into their satisfaction, concerns, and expectations. These reviews represent a rich source of user-generated content that reflects authentic customer perceptions. However, such data are typically unstructured textual information and exist in large volumes, making manual analysis impractical. Therefore, advanced analytical techniques such as text mining and sentiment analysis are required to effectively classify

S. Sodsee et al. (Eds.): IC2IT 2026, LNNS 5590, pp. 56–67, 2026.
https://doi.org/10.1007/978-3-032-28048-0_6

opinions into positive and negative categories and extract meaningful insights from the data.

In the context of pet-friendly hotels, online reviews represent a domain-specific dataset. The experience of staying in pet-friendly accommodations involves unique factors that differ from general hotel services, such as pet policies, additional pet-related fees, pet amenities, cleanliness standards after pet stays, and staff attitudes toward animals. These factors directly influence customer satisfaction and shape the linguistic characteristics of online reviews. Domain-specific vocabulary and contextual expressions such as references to pet odors, designated pet areas, or pet service charges are frequently observed. Consequently, sentiment analysis in this domain requires models capable of capturing contextual nuances specific to pet-friendly hospitality services.

Although sentiment analysis has been widely studied in Thailand, research on pet-friendly hotel reviews is still limited. Most studies focus on general domains, use non-specialized datasets, and rarely compare multiple deep learning models for Thai-language analysis, highlighting a gap in identifying the most suitable techniques for this specific context.

This study aims to address this gap by comparatively evaluating four deep learning models. Recurrent Neural Network, Bidirectional Recurrent Neural Network, Long Short-Term Memory, and Bidirectional Long Short-Term Memory using Thai language online reviews of pet-friendly hotels in the Cha-am and Hua Hin areas collected from Agoda.com. The study provides new insights in three key aspects: (1) the effectiveness of bidirectional architectures in capturing contextual dependencies in domain-specific Thai texts, (2) the performance and generalization capability of different deep learning models when applied to real-world hospitality data, and (3) the practical applicability of the best-performing model within a web-based application to support data-driven decision-making in the hospitality industry.

Deep learning techniques [1] have been widely recognized for their effectiveness in text analysis, particularly sequence-based neural network architectures capable of learning contextual dependencies within textual data. Models such as RNN, Bi-RNN, LSTM, and Bi-LSTM are specifically designed to process sequential information, enabling them to capture both preceding and succeeding contextual information with-in sentences. This capability enhances classification accuracy, especially in languages with complex sentence structures such as Thai.

Applying these techniques to analyze online reviews of pet-friendly hotels in the Hua Hin–Cha-am region is therefore significant both academically and practically. From an academic perspective, it contributes to domain-specific sentiment analysis research in the Thai language. From a practical standpoint, it enables hotel operators to gain deeper insights into customer expectations, improve service strategies, and strengthen their competitive advantage in the growing pet-friendly tourism market.

2 Literature Reviews

2.1 Text Mining

Text mining is the process of transforming unstructured textual data into a structured format in order to identify meaningful patterns and discover new insights. It can be applied to the analysis of large volumes of textual data to capture key concepts, trends, and hidden relationships by employing advanced analytical techniques, such as deep learning. [2].

2.2 Deep Learning Techniques

In this study, deep learning techniques were applied to construct classification models, as described below.

2.2.1 Recurrent Neural Network (RNN)

The Recurrent Neural Network (RNN) is an algorithm well-suited for sequential data, as it can retain information from previous inputs to analyze subsequent data. However, it has limitations when handling long sequences of data [3].

2.2.2 Bidirectional Recurrent Neural Network (Bi-RNN)

The Bidirectional Recurrent Neural Network (Bi-RNN) is an extension of the RNN that processes data in both forward and backward directions. This allows the model to analyze information from both past and future contexts, resulting in a more accurate understanding of textual context [4].

2.2.3 Long Short-Term Memory (LSTM)

Long Short-Term Memory (LSTM) is a type of RNN with a more complex architecture that enables effective learning of long-term dependencies. It is particularly suitable for complex textual data with long-range relationships [5].

2.2.4 Bidirectional Long Short-Term Memory (Bi-LSTM)

Bidirectional Long Short-Term Memory (Bi-LSTM) is an advanced RNN architecture that combines an LSTM with a bidirectional structure. This model can capture long-term dependencies by analyzing textual data in both forward and backward directions, making it well-suited to complex text with distant contextual relationships [6].

2.3 Related Work

Thai word segmentation has reported that Deepcut is an effective deep learning–based approach for word boundary detection. [7] By formulating word segmentation as a character-level binary classification problem using convolutional neural networks, Deepcut demonstrated superior performance compared to dictionary-based and traditional statistical methods. Experimental results across standard Thai corpora showed that Deepcut achieved an F1-score of approximately 96.57%, highlighting its robustness in handling unseen words and complex word structures, and establishing it as a strong baseline for Thai natural language processing tasks.

Comparative analysis of classical machine learning algorithms for sentiment classification in social media text, including Logistic Regression, SVM, Random Forest, Naive Bayes, and XGBoost. [8] The experimental results showed that Logistic Regression and SVM achieved the best performance, with accuracies of 86.22%, indicating their effectiveness for sentiment classification on social media data.

Performed a comparative study of deep learning models for sentiment analysis on social media texts using CNN, CNN-LSTM, and BiLSTM-CNN architectures. [9] The results demonstrated that CNN achieved the highest accuracy of 90.09% on the IMDb dataset, while BiLSTM-CNN performed best on the Sentiment140 dataset with an accuracy of 82.10%, highlighting the suitability of CNN-based and hybrid architectures for sentiment analysis.

Proposed a deep learning approach for sentiment analysis of online reviews using word embedding techniques combined with CNN, LSTM, and Bi-LSTM models. [10] The study revealed that Bi-LSTM with GloVe embeddings achieved the best performance, obtaining 95.75% accuracy, precision, recall, and F1-score for binary sentiment classification.

Comprehensive review of deep learning models for Twitter sentiment analysis, covering DNN, CNN, RNN, LSTM, Bi-LSTM, GRU, and BERT architectures. [11] The study concluded that BERT and Bi-LSTM consistently delivered superior performance, especially for short and informal Twitter texts, while also discussing key challenges and future research opportunities.

Investigated the distinctive attributes of pet-friendly hotels and their influence on customer behavior using both qualitative and quantitative approaches. [12] Importance–performance analysis and regression analysis were employed to examine consumer responses. The study identified six key attributes, particularly pet-related activities, which significantly influence customers' intention to use services and engage in word-of-mouth communication. In addition, pet-oriented facilities were found to increase customers' willingness to pay a premium, providing strategic insights for positioning hotels as pet-friendly brands.

Developed a deep learning–based sentiment analysis system to evaluate hotel service quality within the context of sustainable tourism. [13] The study analyzed 15,522 customer reviews of five-star hotels in Antalya, Turkey, using text mining, topic classification, and sentiment analysis. The results revealed that Hotel HB2 achieved the highest F1-score of 97.9%, while overall customer satisfaction reached 91%, and the emotional dimension accounted for 77%. Key influencing factors included cleanliness, food quality, and staff performance. In addition, customer reviews reflected hotels' environmental responsibility, which significantly impacted customer satisfaction. The findings demonstrate that AI-based techniques can effectively analyze large-scale review data. Consequently, this study provides practical insights for hotel managers to enhance service quality, manage customer loyalty, and develop sustainable strategic initiatives.

3 Research Methodology

This research aims to compare the performance of sentiment analysis techniques on user opinions regarding pet-friendly hotels on social media platforms, as illustrated in the conceptual framework shown in Fig. 1.

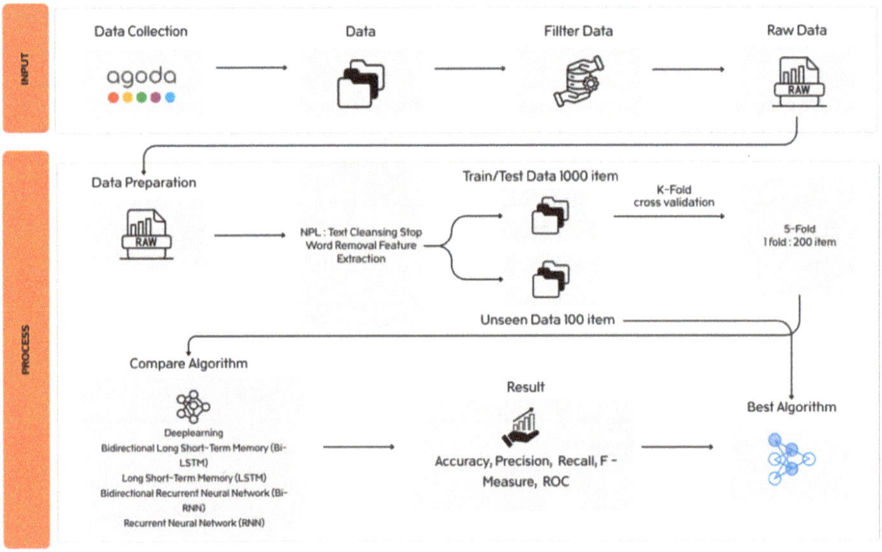

Fig. 1. Conceptual framework for research.

As shown in Fig. 1 the conceptual framework of this study consists of two main components: data collection and data preprocessing, and performance comparison.

3.1 Collect Data

User reviews of hotel services were collected from the Agoda.com website from January 2024 to June 2025, totaling 1,000 textual reviews. The reviews were obtained from online platforms and categorized into positive and negative sentiments. The collected data were stored in Excel file format, after which the reviews were carefully selected and manually labeled according to their sentiment polarity.

Subsequently, all reviews were processed through a text filtering and data preparation procedure, which consisted of several steps. First, symbols and special characters were removed from the text. Next, tokenization was performed to segment the text into individual lexical units. Then, stop-word removal was applied to eliminate in-significant words while preserving the original meaning of the text. After that, feature extraction was conducted to identify important features and transform the data into a format suitable for computational processing. This process ensured that the dataset was properly prepared for use in the model development phase. Details of sample reviews are presented in Table 1.

Table 1. Sample user reviews.

Hotel Review	Type
This location is excellent for a holiday stay and for visiting nearby local markets. The swimming pool and facilities are suitable for families; however, in December, the water may be slightly too cold	Positive
The accommodation and hotel are very clean	Positive
The rooms are spacious, and the staff provide excellent service	Positive
The poolside staff were impolite. The hotel lacks luxury relative to its price. The corridors are maze-like, and the elevators require long waiting times	Negative
The hotel is old, overpriced, noisy due to the air conditioner, and unclean	Negative
The room price was not worth the cost. The room had a musty odor and appeared outdated. Although the stay was offered at a discounted rate, expectations were higher for a room priced at 3,000 baht rather than approximately 1,000 baht. If calculated at the full price prior to the discount, the value was even less satisfactory	Negative

Based on Table 1 the data were processed in a data preparation phase prior to model comparison. This process consists of the following steps.

3.1.1 Data Cleaning

Data cleaning involves improving data quality by removing special characters, eliminating extra spaces, deleting line breaks, and discarding irrelevant symbols.

3.1.2 Word Segmentation

Word segmentation is the process of dividing text into lexical units using the Deep Cut technique.

3.1.3 Model Performance Evaluation

The performance of the models was evaluated using stratified k-fold cross-validation with k = 5 folds. This method partitions the dataset into five subsets, where each fold maintains a class distribution similar to that of the entire dataset. This approach helps reduce potential bias arising from random data splitting and enhances the reliability of the evaluation results. In each iteration, four folds were used for model training, and the remaining fold for testing. This procedure was repeated until all five folds had been used as the test set, and the results from all folds were averaged to represent the overall performance of the models.

3.2 Hyperparameters Setup

The models were compared using four deep learning techniques: Recurrent Neural Networks (RNNs), Bidirectional Recurrent Neural Networks (Bi-RNNs), Long Short-Term Memory (LSTMs), and Bidirectional Long Short-Term Memory (Bi-LSTMs). The model parameters were tuned for comparison, as presented in the table.

Table 2. Parameter settings used for model tuning.

Parameter	Value
Memory Units	32, 64, 128, 256
Dense Units	32, 64, 128, 256
Dropout Rate	0.1, 0.2, 0.3, 0.4, 0.5, 0.6, 0.7
Epochs	10,30, 50,70,90

Based on Table 2 a total of 311 parameter configurations were evaluated. The memory units were set to [32, 64, 128, 256][32, 64, 128, 256][32, 64, 128, 256], the dense units were set to [32, 64, 128, 256][32, 64, 128, 256][32, 64, 128, 256], the number of epochs was set to [10, 30, 50, 70, 90][10, 30, 50, 70, 90][10, 30, 50, 70, 90], and the dropout rate was set to [0.1, 0.2, 0.3, 0.4, 0.5, 0.6, 0.7][0.1, 0.2, 0.3, 0.4, 0.5, 0.6, 0.7][0.1, 0.2, 0.3, 0.4, 0.5, 0.6, 0.7].

4 Experimental Result

The model performance was evaluated using accuracy (A), precision (P), recall (R), F1-score (F), and Receiver Operating Characteristic (ROC) analysis. The experimental results are presented in Tables 3, 4, 5, 6, 7 and 9.

Table 3. Results of the Bi-LSTM model.

Parameter				Performance				
Unit	DenseUnit	Dropout	Epoch	A(%)	P	R	F	ROC
64	**64**	**0.2**	**50**	**85.40**	**0.857**	**0.854**	**0.854**	**0.918**
64	64	0.7	70	85.40	0.856	0.854	0.854	0.918
256	128	0.3	70	85.30	0.854	0.853	0.853	0.918
32	32	0.2	50	85.30	0.854	0.853	0.853	0.913
32	32	0.7	70	85.30	0.854	0.853	0.853	0.913
256	64	0.7	50	85.30	0.854	0.853	0.853	0.916
128	128	0.3	50	85.20	0.853	0.852	0.852	0.916
128	64	0.1	10	85.20	0.853	0.852	0.852	0.917
32	64	0.7	90	85.20	0.853	0.852	0.852	0.918
128	32	0.4	30	85.20	0.854	0.852	0.852	0.913

Based on Table 3 the Bi-LSTM model with 64 Bi-LSTM units, 64 dense units, a dropout rate of 0.2, and 50 epochs achieved the best performance, with an accuracy of 85.40%.

Table 4. Results of the LSTM model.

Parameter				Performance				
Unit	Dense Unit	Dropout	Epoch	A(%)	P	R	F	ROC
256	**128**	**0.6**	**30**	**60.30**	**0.626**	**0.603**	**0.517**	**0.92**
32	64	0.1	50	58.70	0.592	0.587	0.511	0.92
128	128	0.1	30	58.50	0.714	0.585	0.504	0.92
256	256	0.6	50	58.00	0.646	0.580	0.470	0.91
32	32	0.4	50	57.90	0.713	0.579	0.495	0.91
128	128	0.5	30	57.70	0.618	0.577	0.495	0.92
64	128	0.5	50	57.70	0.583	0.577	0.498	0.92
256	128	0.6	50	57.60	0.721	0.576	0.484	0.92
256	128	0.5	90	57.60	0.548	0.576	0.464	0.92
64	128	0.6	90	57.50	0.612	0.575	0.496	0.91

Based on Table 4 the LSTM model with 256 LSTM units, 128 dense units, a dropout rate of 0.6, and 30 epochs achieved the best performance, with an accuracy of 60.30%.

Table 5. Results of the LSTM model.

Parameter				Performance				
Unit	Dense Unit	Dropout	Epoch	A(%)	P	R	F	ROC
64	**32**	**0.2**	**50**	**82.70**	**0.831**	**0.827**	**0.827**	**0.920**
32	64	0.2	50	81.40	0.817	0.814	0.814	0.918
64	128	0.4	30	81.10	0.813	0.811	0.811	0.918
32	128	0.1	50	80.70	0.812	0.807	0.806	0.918
128	128	0.5	50	80.60	0.813	0.806	0.804	0.913
32	128	0.3	30	80.50	0.811	0.805	0.804	0.913
64	64	0.3	50	80.40	0.808	0.804	0.803	0.916
32	128	0.4	10	80.30	0.804	0.803	0.803	0.916
64	128	0.2	30	80.20	0.806	0.802	0.801	0.917
64	128	0.1	50	80.20%	0.806	0.802	0.801	0.918

Based on Table 5 the Bi-RNN model with 64 Bi-RNN units, 32 dense units, a dropout rate of 0.2, and 50 epochs achieved the best performance, with an accuracy of 82.70%.

Table 6. Results of the RNN model.

Parameter				Performance				
Unit	Dense Unit	Dropout	Epoch	A(%)	P	R	F	ROC
32	32	0.3	10	71.40	0.721	0.714	0.711	0.918
32	64	0.3	10	69.30	0.705	0.693	0.687	0.918
32	32	0.1	50	68.60	0.638	0.686	0.652	0.918
32	64	0.2	10	67.90	0.688	0.679	0.673	0.913
32	128	0.3	50	67.50	0.686	0.675	0.668	0.913
32	64	0.4	50	66.30	0.664	0.663	0.662	0.916
32	128	0.4	50	66.00	0.665	0.660	0.656	0.916
32	64	0.1	50	66.00	0.662	0.660	0.658	0.917
32	32	0.6	70	65.90	0.671	0.659	0.648	0.918
32	32	0.1	10	65.70	0.659	0.657	0.655	0.913

Based on Table 6 the RNN model with 32 RNN units, 32 dense units, a dropout rate of 0.3, and 10 epochs achieved the best performance, with an accuracy of 71.40%.

Table 7. Performance evaluation results of the models using unseen data.

Model	Performance				
	A(%)	P	R	F	ROC
Bi-LSTM	**0.71**	**0.67**	**0.71**	**0.67**	**0.71**
LSTM	0.50	0.25	0.50	0.33	0.49
Bi-RNN	0.63	0.68	0.63	0.60	0.57
RNN	0.59	0.57	0.59	0.54	0.62

Based on Table 7 the models were evaluated using the same unseen dataset consisting of 100 samples. The results indicate that the Bi-LSTM model achieved the highest performance, with an accuracy of 86.70%, an average precision of 0.87, an average recall of 0.87, and an average F1-score of 0.87.

Table 8. Example of sentiment analysis output (example 1).

Item	Result	
Text	Hong Yai Kwang khwang	
Analysis	Positive	
Accuracy	68.31%	
Category	Keywords	Status
Service	3	YES
Room	0	NO
Location	0	NO
Food	0	NO

Based on Tables 8. and 9 the sentiment analysis results produced by the Bi-LSTM model with 64 Bi-LSTM units, 64 dense units, a dropout rate of 0.2, and 50 epochs demonstrate effective performance in sentiment classification, as presented in the tables.

Table 9. Example of sentiment analysis output (example 2).

Item	Result	
Text	Hongnam Sokkaprok Mak KlinMen Mak	
Analysis	Negative	
Accuracy	77.07%	
Category	Keywords	Status
Service	0	NO
Room	2	YES
Location	0	NO
Food	0	NO

5 Conclusion

This study analyzes user sentiment toward pet-friendly hotels on online platforms using deep learning techniques. A total of 1,000 reviews were collected from June 2024 to June 2025 and classified into two categories: positive and negative sentiment. After data selection, the dataset was processed through a data preparation pipe-line comprising data cleaning, word segmentation using the DeepCut technique, text-to-sequence numerical encoding, and numerical encoding of sentiment labels. To address class imbalance, the class weight technique was applied. The dataset was then evaluated using Stratified K-Fold Cross-Validation with $k = 5$, in which the data were partitioned into five subsets.

Subsequently, four deep learning models Bidirectional Long Short-Term Memory (Bi-LSTM), Long Short-Term Memory (LSTM), Bidirectional Recurrent Neural Network (Bi-RNN), and Recurrent Neural Network (RNN) were developed, and their parameters were tuned. Model performance was assessed using accuracy, precision, recall, F1-score, and Receiver Operating Characteristic (ROC) analysis. The results indicate that the Bi-LSTM model achieved the best overall performance, with an accuracy of 85.40%, a precision of. This superior performance can be attributed to the bidirectional architecture of the Bi-LSTM model, which enables it to capture contextual information from both forward and backward directions. As a result, the model can more effectively understand long-range dependencies and semantic relationships within textual data compared to unidirectional models.

References

1. Chitboonyapinij, C., Maneerat, P., Jirawichitchai, N.: Development of a sentiment analysis model for Thai social media using deep learning techniques. Huachiew Chalermprakiet Univ. J. Sci. Technol. **8**(2), 8–18 (2022)
2. IBM: What is text mining? IBM Think. http://www.ibm.com/think/topics/text-mining. Accessed 12 Dec 2024

3. Mienye, I.D., Swart, T.G., Obaido, G.: Recurrent neural networks: a comprehensive review of architectures, variants, and applications. Information **15**(9), 517 (2024)
4. Chowdhury, S., et al.: A multitask bi-directional RNN model for named entity recognition on Chinese electronic medical records. BMC Bioinform. **19**(Suppl 17), 499 (2018)
5. Kilinc, M.: LSTM-based time series forecasting of user-derived quality signals in mobile banking systems. Systems **13**(11), 949 (2025)
6. GeeksforGeeks: Bidirectional LSTM in NLP. https://www.geeksforgeeks.org/bidirectional-lstm-in-nlp/. Accessed 20 Feb 2024
7. Phatthiyaphaibun, W., Boonkwan, P.: Thai word segmentation using deep learning techniques: a review. J. Electr. Comput. Eng. (2023). Article ID 8592214
8. Jahan, et al.: Comparative analysis of machine learning algorithms for sentiment classification in social media text. World J. Adv. Res. Rev. **23**(3), 2842–2852 (2024)
9. Derbentsev, V.D., et al.: A comparative study of deep learning models for sentiment analysis of social media texts. In: CEUR Workshop Proceedings, vol. 3485, pp. 168–188 (2023)
10. AlSurayyi, W.I., Alghamdi, N.S., Abraham, A.: Deep learning with word embedding modeling for sentiment analysis of online reviews. Int. J. Comput. Inf. Syst. Ind. Manage. Appl. **11**, 227–241 (2019)
11. Chaudhary, L., et al.: A review of deep learning models for Twitter sentiment analysis: challenges and opportunities. IEEE Trans. Comput. Soc. Syst. **11**(3), 3550–3579 (2024)
12. Kim, J., et al.: Pet-friendly hotels: distinctive attributes and their influence on customer behavior. J. Hospitality Tourism Insights **8**(6), 2359–2380 (2025)
13. Erdoğan, D., et al.: Developing a deep learning-based sentiment analysis system of hotel customer reviews for sustainable tourism. Sustainability **17**(13), 5756 (2025)

Development of a Decision Support System for Housing Purchase or Rental Selection Using Data Mining Techniques

Suradech Leerujikul(✉) and Watchareewan Jitsakul

Faculty of Information Technology and Digital Innovation, King Mongkut's University of Technology North Bangkok, Bangkok, Thailand
s6707021857015@email.kmutnb.ac.th,
watchareewan.j@itd.kmutnb.ac.th

Abstract. This study compares the performance of machine learning algorithms CatBoost, Convolutional Neural Network (CNN), and Multinomial Naive Bayes for developing a decision support system for housing purchase or rental selection in Bangkok and its metropolitan area. Recursive Feature Elimination (RFE) was applied to identify influential factors and reduce model complexity. Experimental results show that CatBoost achieved the highest performance, with an accuracy of 0.83 using all features, and maintained competitive performance when reduced to seven selected features. CNN demonstrated moderate predictive capability, while Multinomial Naive Bayes showed the lowest performance across all feature subsets. Feature importance analysis revealed that factors such as age, monthly income, gender, and housing status significantly influence housing decisions. The findings indicate that CatBoost combined with RFE is an effective approach for developing a reliable housing decision support system under current economic and social conditions.

Keywords: Machine Learning · CatBoost · Recursive Feature Elimination · Decision Support System · Housing

1 Introduction

Thailand is rapidly becoming a super-aged society, driven by declining birth rates with Generation Z and Alpha accounting for only 20.23% and 11.28% of the population, respectively. This demographic shift, coupled with persistent inflation, places immense financial pressure on the working-age population. The rising cost of living continues to erode real income and weaken overall financial resilience in this challenging economic climate.

The fragile economic environment, characterized by income instability and limited financial literacy, has weakened long-term financial security for much of the population. These challenges are acute in the housing sector, where property prices rise by 5–10% annually. Consequently, households face severe affordability constraints and lack the

S. Sodsee et al. (Eds.): IC2IT 2026, LNNS 5590, pp. 68–77, 2026.
https://doi.org/10.1007/978-3-032-28048-0_7

confidence to commit to long-term mortgage obligations amidst rising costs and stagnant earnings.

Amidst economic uncertainty, housing decisions have become increasingly complex, requiring a balance of financial capacity, employment stability, and long-term needs. While homeownership offers stability, it entails significant financial risks. In contrast, renting provides flexibility but lacks long-term security. These critical trade-offs make residential decision-making exceptionally challenging in the current socio-economic climate.

Although prior studies have examined housing affordability and residential choice in the Bangkok Metropolitan Region, such as the work of Poolsawat et al. (2025) which identified that age and monthly income are predominant factors influencing purchase decisions [1], many still rely on traditional statistical methods or focus on limited socioeconomic variables, providing limited decision support in real-world contexts. Moreover, existing systems often fail to integrate advanced machine learning techniques with systematic feature selection to enhance interpretability and predictive accuracy. Consequently, there remains a research gap in the development of intelligent decision support systems that can effectively assist individuals in making purchase-or-rent housing decisions under complex economic conditions.

To address this gap, this study proposes a decision support system for residential purchase-or-rent decisions in the Bangkok Metropolitan Region and its surrounding areas. The proposed framework integrates data mining techniques with machine learning algorithms, including CatBoost [2], Multinomial Naive Bayes [3], and Convolutional Neural Networks (CNN) [4]. Recursive Feature Elimination is employed to identify influential decision factors and improve model performance. Furthermore, the system's effectiveness and user satisfaction are evaluated to ensure its practical applicability. The findings of this study aim to provide a robust, data-driven tool to support housing decisions aligned with the current economic and social context of the study area. Literature Reviews.

1.1 Diverse Counterfactual Explanations

Diverse Counterfactual Explanations (DiCE), proposed by Mothilal et al. (2020), is a framework for enhancing the interpretability of machine learning models by generating a set of diverse hypothetical scenarios that lead to a desired outcome. Unlike traditional counterfactual methods that often provide a single solution, DiCE formulates the generation process as an optimization problem that balances three primary objectives: validity, proximity, and diversity. The framework ensures validity by requiring that all generated counterfactuals belong to the target class, while proximity minimizes the distance between the original input and the counterfactuals to ensure subtle changes. Crucially, DiCE incorporates a diversity constraint via a Determinantal Point Process (DPP), encouraging the generation of counterfactuals that are distinct in the feature space. This approach provides users with a broader range of actionable options that account for various real-world constraints and preferences, thereby improving the transparency and utility of Explainable AI (XAI) systems [5].

1.2 Related Studies

Krit [6] investigated early-stage plant stress detection using deep learning implemented on an FPGA to enhance future agricultural yields. The study used CNN and hybrid CNN + LSTM models trained on a dataset of over 40,000 multi-view images of sorghum plants under three nutrient-deficiency conditions. When implementing the CNN model on the PYNQ-Z1 (FPGA) board, the results showed a maximum accuracy of only 61%, indicating significant underfitting due to hardware resource constraints. Despite the FPGA processing being 38 times faster than a conventional CPU, its accuracy was insufficient for practical applications.

In contrast, the VGG16 + LSTM model achieved 85% accuracy within 3 days of nutrient deprivation, rising to 91% by day 13. This hybrid model demonstrated a superior ability to learn temporal features compared to standalone CNNs. Although its processing time was 20.72 times longer, it proved highly suitable for non-real-time applications. The findings suggest that sequential models, such as LSTMs, are more effective at detecting early physical changes in plants. Future research is recommended to focus on model scaling for embedded systems and expanding datasets to include other plant species and various stress conditions.

Pawaris [7] conducted a study on developing a prototype for predicting travel duration on Bangkok's roads using machine learning. The research utilized Mobile Probe data from the iTIC Foundation, encompassing 4,634 vehicles. This data was transformed into Origin-Destination (OD) format with 29 variables, including coordinates, time, administrative districts, and weather conditions.

The data preprocessing involved rigorous cleaning and the application of K-Fold Cross-Validation ($K = 5$) alongside 100 iterations of RandomizedSearchCV for hyperparameter tuning. The results indicated that the Random Forest model achieved the highest performance (MAE = 134.72, MAPE = 43.06%), although it was computationally intensive. In contrast, XGBoost and LightGBM yielded comparable performance while significantly reducing processing time. Notably, while CatBoost showed lower overall accuracy (MAE = 148.41, MAPE = 52.55%), its training time was substantially shorter than the other models, highlighting its potential for applications requiring rapid model training. The study suggests that future research should focus on further hyperparameter optimization and updating datasets to the post-COVID-19 era to enhance the model's predictive accuracy.

Adnan et al. [8] conducted a Systematic Literature Review on the RFE algorithm, analyzing 76 peer-reviewed articles published between 2018 and 2023 to evaluate its evolution and applications. Their findings categorize RFE implementations into three primary domains: Standard RFE (49%), Strategy-based or Hybrid RFE (43%), and Modified RFE (8%), highlighting a significant shift toward integrating RFE with complementary feature selection techniques to optimize performance. The study further identifies that RFE is most frequently paired with Support Vector Machines (SVM) at 19.5% and Random Forest (RF) at 16.7%, underscoring its robustness in handling high-dimensional data across various classification and prediction tasks. Ultimately, the research emphasizes that current trends are moving toward hybrid frameworks to address computational complexity while enhancing feature subset accuracy, providing a critical foundation for future developments in machine learning-based dimensionality reduction.

Prokhorenkova et al. [9] introduced CatBoost, a Gradient Boosting Decision Tree (GBDT) framework designed to eliminate internal biases in traditional models. The study addresses Prediction Shift and Categorical Feature complexities through the implementation of Ordered Boosting and Ordered Target Statistics (Ordered TS), preventing target leakage while enhancing generalization. Experimental results demonstrate that CatBoost consistently outperforms state-of-the-art models like XGBoost and LightGBM in predictive accuracy across various benchmarks, providing a robust foundation for complex data mining tasks.

2 Research Methodology

The system development methodology began with a content validity assessment of the research instrument using the IOC, verified by three experts. Data were collected via Google Forms from a sample population in the Bangkok Metropolitan Region. The preprocessing phase involved addressing class imbalance through data resampling and performing feature selection to optimize model performance. Multiple machine learning algorithms were benchmarked to identify the optimal predictive model. The final model was implemented into a prototype system using a Python-based backend and a React-based user interface. Furthermore, the system incorporates DiCE to provide actionable insights. When the model predicts a lack of purchase readiness, DiCE generates targeted parameter adjustments to help users improve their future homeownership eligibility.

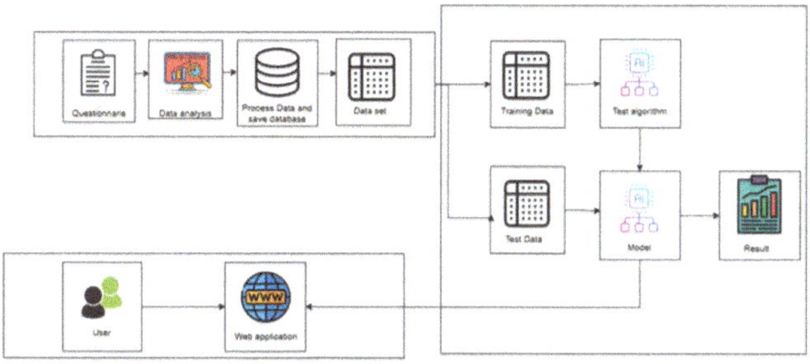

Fig. 1. Conceptual framework for research.

Figure 1 the framework illustrates three primary phases of the study: data collection via questionnaires followed by data analysis to establish the dataset; the processing phase, which integrates RFE for feature selection and model training using algorithms such as CatBoost and CNN and the final deployment phase as a web application. This system serves as a decision support tool, enabling users to receive data-driven predictions for residential purchase-or-rent decisions amidst current economic challenges.

2.1 Collect Data

The research utilized a 28-item questionnaire covering demographics, financial security, and residential attitudes. Data were collected via Google Forms using random sampling across the Bangkok Metropolitan Region, resulting in 1,107 respondents (379 homeowners and 728 non-homeowners). To mitigate class imbalance and prevent model bias, a data balancing technique was applied, resulting in a final dataset of 1,456 samples for model training and evaluation.

2.2 Hyperparameters Setup

For model development, three machine learning algorithms were selected: CatBoost, CNN, and Multinomial Naive Bayes. To ensure a robust and rigorous evaluation, the dataset was partitioned into a training set (80%) and a test set (20%), where the training portion was utilized for model construction and hyperparameter tuning through 10-fold cross-validation (k = 10). This process involved dividing the training data into ten subsets, rotating the training and validation phases to ensure the models were well-generalized and to minimize the risk of overfitting. The initial results indicated that CatBoost was the most effective predictive model; consequently, it was employed for feature selection via RFE:

$$J_i = \Sigma_{t=1}^{T} \Sigma_{n \in S_{i,t}} \left(\left(v_1^n - \overline{v}^{(n)} \right)^2 \cdot w_1^n + \left(v_2^n - \overline{v}^{(n)} \right)^2 cdotw_2^n \right) \quad (1)$$

a wrapper-based technique that identifies the most significant variables. This process yielded the Top 7 and Top 10 feature sets, including key variables such as age, monthly income, and mortgage repayment readiness. Subsequently, these feature sets were re-evaluated using all three algorithms with further hyperparameter optimization. Finally (Fig. 2),

Fig. 2. Workflow of Model Selection, Feature Selection, and Retraining Process.

3 Experimental Results

This research comparatively evaluated the learning performance of three algorithms: CatBoost, CNN, and Multinomial Naive Bayes, alongside the implementation of RFE. Detailed experimental results, including model performance metrics and feature rankings, are presented in Tables 1, 2, 3, 4, 5 and 6.

Table 1. Report CatBoost Result.

Measure	All Feature	10 Features	7 Features
Accurcy	0.8	0.83	0.82
Precision	0.8	0.83	0.82
Recall	0.79	0.82	0.81
F1	0.8	0.82	0.81
ROC	0.88	0.89	0.86

Table 1 presents results from the unseen 20% test set, ensuring an unbiased evaluation of model generalization. The Top 7 feature subset was selected for final implementation as it provides an optimal balance between predictive accuracy and reduced complexity, with CatBoost maintaining strong performance despite fewer variables.

Table 2. Report CatBoost Best Parameter.

Hyperparameter	All Feature	10 Features	7 Features
Learning Rate	0.03	0.03	0.03
Iterations	500	500	800
Depth	6	6	5
L2 Leaf Reg	5	30	50
Random Strength	20	30	20
Bagging Temperature	0.5	0.5	1
Border Count	50	50	32

Table 2 presents the experimental results, detailing the optimal parameters (Best Parameters) obtained through the systematic experimentation and hyperparameter tuning process.

Table 3. Report CNN Result.

Measure	All Feature	10 Features	7 Features
Accurcy	0.76	0.69	0.70

(continued)

Table 3. *(continued)*

Measure	All Feature	10 Features	7 Features
Precision	0.78	0.69	0.70
Recall	0.76	0.69	0.70
F1	0.79	0.68	0.67
ROC	0.84	0.77	0.77

Table 3 reveals that while using all features yields the best results for CNN, the CatBoost model still maintains superior performance across all configurations. Furthermore, reducing the feature count to 10 and 7 leads to a significant decline in CNN's efficiency compared to the full feature set.

Table 4. Report Multinomial Naive Result.

Measure	All Feature	10 Features	7 Features
Accurcy	0.67	0.67	0.67
Precision	0.67	0.67	0.67
Recall	0.66	0.66	0.66
F1	0.66	0.66	0.66
ROC	0.75	0.75	0.75

Table 4 details the optimal hyperparameters for the model. The experimental results indicate that across all scenario including the full feature set, the Top 10 features, and the Top 7 feature the optimal alpha parameter remained consistently at 0.0001.

Table 5. Report CNN Best Parameter.

Hyperparameter	All Feature	10 Features	7 Features
Conv1 Filters	32	32	32
Add Conv2	True	True	True
Dense Units	96	96	64
Dropout	0.3	0.00	0.3
Learning Rate	0.00151	0.00051	0.00101
Conv2 Filters	96	128	96
Kernel Size (Conv2)	3	6	3

Table 5 according to Tables 3 and 5, the CNN model achieves peak performance across all metrics (Accuracy 0.7) when utilizing all features with its most complex hyperparameter configuration. While feature reduction simplifies model parameters, it leads to a significant performance drop compared to the full feature set; furthermore, CatBoost consistently outperforms CNN across all experimental scenarios.

Table 6. Report RFE score.

Feature	CatBoost Importance	RFE_Score
age	11.71124	24
Monthly Income	7.95124	24
Weekly Commuting Days	6.48124	24
Residential Area	7.82223	26
Gender	13.97	22
Home Purchase Status	12.54	19
Current Homeownership Status	12.54	15
Employment Status	12.04	12
Marital Status	11.02	6
Job Change Frequency	7.69	3

Table 6 illustrates the feature importance rankings based on CatBoost Importance and RFE Scores. The data reveals that Gender, Home Purchase Status, and Age are the most significant predictors, whereas Job Change Frequency exhibits the lowest importance across both evaluation metrics. These findings carry significant practical and economic implications; for instance, Age and Home Purchase Status often serve as proxies for financial stability and life stages within housing economics. According to the Life-Cycle Hypothesis, individuals in different age cohorts possess varying levels of accumulated wealth and demands for residential stability, directly influencing their propensity to purchase rather than rent. From a policy perspective, understanding that these socio-economic factors dominate the decision-making process allows for the development of more targeted financial products and housing assistance programs. For users, this ranking enhances transparency by identifying the primary drivers of their housing recommendations, providing a "financial reality check" that aligns with their current life stage and economic capacity.

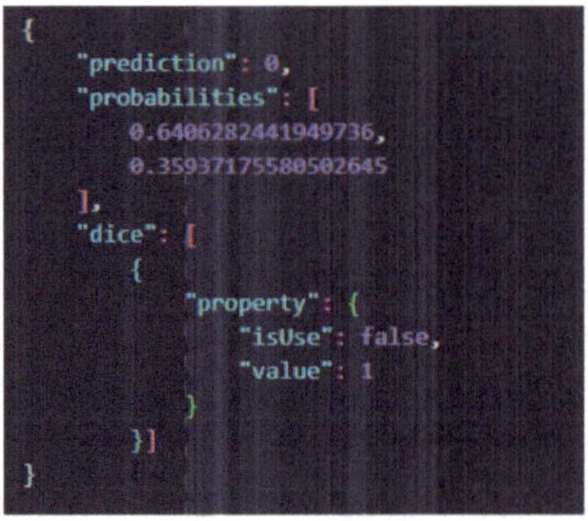

Fig. 3. Response API.

Figure 3 illustrates the system's API response, where a prediction of "0" suggests "Rent." To offer actionable insights, the system utilizes DiCE with domain-specific constraints: fixed demographic features (e.g., sex), while socio-economic factors (e.g., income, age) are restricted to increasing values, and commuting expenses are allowed to decrease. This constrained optimization ensures that generated counterfactuals provide feasible paths for users to reach a "Purchase" outcome, enhancing model transparency and practical utility.

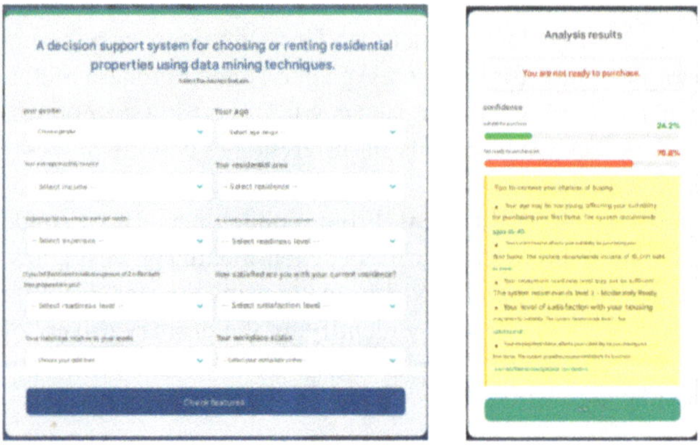

Fig. 4. Web application.

Figure 4 the web-based User Interface of the decision support system developed using React. The interface consists of two primary components: (a) the input form, which features a comprehensive set of personal, financial, and residential factors for data collection, and (b) the output display, which presents the user's homeownership readiness status, confidence scores, and "Recommendations to Improve Purchase Opportunities." These recommendations utilize model-driven insights and DiCE to suggest actionable adjustments, such as changes in income or employment status, providing clear guidance for user decision-making.

4 Conclusion

This study found that the CatBoost algorithm achieved the highest accuracy among all evaluated models, as shown in Table 1. Based on this result, the CatBoost model was further integrated with RFE to identify the most influential features affecting model performance. The selected features were subsequently evaluated, and the experimental results are reported in Tables 1, 3, and 5. According to the findings, the CatBoost model using seven selected features was chosen for system development due to its strong predictive performance and reduced model complexity. However, this study is subject to certain limitations, as the available data were relatively small and lacked sufficient variability due to time and budget constraints. Therefore, future studies are recommended to consider longer data-collection periods and increased resources to obtain larger, more diverse datasets, which could further improve model robustness and generalization performance.

References

1. Poolsawat, K., et al.: A study of housing purchase decision-making in bangkok and metropolitan areas. In: The 15th BENJAMIT Network National & International Conference, pp. 137–147. Thonburi University, Bangkok (2025)
2. Kumar, R., Singh, R.: Comparison of the CatBoost classifier with other machine learning models. In: Proceedings of the 2020 International Conference on Computer Science, Engineering and Applications (ICCSEA) (2020)
3. Abbas, M., Ali, K.A., Jamun, S., Memon, J.: Multinomial Naive Bayes classification model for sentiment analysis. Int. J. Comput. Sci. Netw. Secur. (IJCSNS) **19**(3), 62–67 (2019)
4. Singh, A., Kumar, R.: Heart disease prediction using machine learning algorithms. In: Proceedings of the 2020 International Conference on Electrical and Electronics Engineering (ICE3), pp. 452–457 (2020)
5. Mothilal, R.K., Sharma, A., Tan, C.: Explaining machine learning classifiers through diverse counterfactual explanations. In: Proceedings of the ACM Conference on Fairness, Accountability, and Transparency (FAccT), pp. 607–617 (2020)
6. Rojrungruangporn, K.: Early-stage plant stress detection using deep learning developed on FPGA. Thesis/Report, Faculty of Engineering, Chulalongkorn University (2022)
7. Wetchawanakitkul, P.: Forecasting gold price direction based on sentiment analysis of Twitter texts using machine learning techniques. Thesis/Report, Faculty of Engineering, Chulalongkorn University (2022)
8. Adnan, N., Islam, M.Z.: A systematic literature review of recursive feature elimination algorithms. ResearchGate (2024)
9. Prokhorenkova, L., et al.: CatBoost: unbiased boosting with categorical features. In: Advances in Neural Information Processing Systems (2018)

End-to-End Multimodal Accident Risk Prediction Pipeline Integrating Deep Learning, Risk Fusion, and LLM-Powered Voice Alerts

Harshita Pendli(✉) and Srinivas Naik Nenavath

Department of Computer Science and Engineering, Indian Institute of Information Technology Design and Manufacturing (IIITDM), Kurnool, Andhra Pradesh, India
harshithaapple4@gmail.com, srinu@iiitk.ac.in

Abstract. Accident risk prediction is essential for improving safety in intelligent transportation systems under complex real-world driving conditions influenced by driver behaviour, weather, and road quality. This paper presents a validated end-to-end multimodal accident-risk prediction framework integrating driver drowsiness detection (MobileNetV3-Large), weather classification (EfficientNet-B0), and road-hazard detection (YOLOv8n) into a unified risk assessment pipeline. To enhance interpretability, a locally deployed large language model generates concise, context-aware alerts synchronized with multimodal video streams in an offline dashboard. Extensive experiments with multiple datasets and baseline models are evaluated using accuracy, precision, recall, F1-score, and mAP@0.5 at both module and system levels. Results demonstrate that the proposed multimodal fusion framework achieves 92% overall system-level risk prediction accuracy while providing interpretable alerts, highlighting its practical potential for intelligent driver-assistance and accident prevention systems.

Keywords: Accident Risk Prediction · Multimodal Fusion · Driver Drowsiness Detection · Weather Classification · Road Hazard Detection · YOLOv8n · MobileNetV3-Large · EfficientNet-B0 · Large Language Models

1 Introduction

Road accidents continue to be a major global safety issue, often resulting from the combined influence of driver fatigue, adverse weather, and poor road conditions. Most existing driver-assistance systems focus on a single risk factor, such as drowsiness detection or road anomaly detection, and therefore fail to capture the overall driving context. Recent progress in deep learning and computer vision has made it possible to analyse driver behaviour, environmental conditions, and road quality simultaneously for accident-risk prediction [1–4].

S. Sodsee et al. (Eds.): IC2IT 2026, LNNS 5590, pp. 78–87, 2026.
https://doi.org/10.1007/978-3-032-28048-0_8

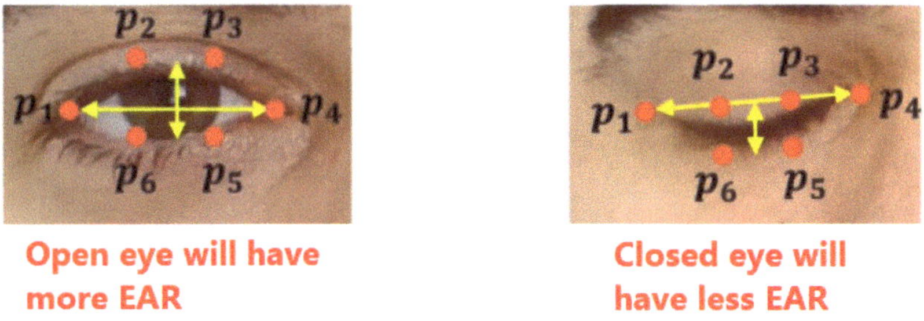

Fig. 1. Eye Aspect Ratio (EAR) computation from facial landmarks.

However, integrating these heterogeneous risk factors into a unified and interpretable framework suitable for practical driving scenarios remains a challenging task. Moreover, several prior studies evaluate perception modules independently rather than assessing the performance of a complete end-to-end multimodal system, leading to solutions that are closer to feasibility demonstrations than deployable safety frameworks. To overcome these limitations, this paper presents an end-to-end multimodal accident-risk prediction framework that jointly analyses driver state, weather conditions, and road hazards within a single pipeline [5–8]. The confidence outputs from these modules are fused using a weighted risk fusion strategy to compute an overall risk score and corresponding risk level. To improve transparency, the fused risk information is further processed by a locally deployed large language model (Ollama Mistral 7B) to generate short, context-aware alerts that explain the detected risk factors. These alerts are synchronized with driver, weather, and road video streams using the MoviePy framework to provide an interpretable offline dashboard visualization [9,10]. In contrast to single-modality systems, the proposed approach emphasizes both module-level and system-level evaluation using quantitative metrics and baseline comparisons to validate the effectiveness of multimodal fusion for accident-risk prediction. The main contribution of this work is the development of an interpretable and integrated multimodal framework that combines deep-learning-based perception, risk fusion, and language-based explanation to deliver reliable and practically relevant driving-risk assessment for intelligent driver-assistance applications [10] (Fig. 1).

2 Literature Survey

Recent research has widely explored the use of computer vision and deep learning to enhance road safety and intelligent transportation systems [1–4]. Most existing studies focus on isolated risk factors such as driver behaviour monitoring, weather recognition, or road-hazard detection. While these approaches have achieved promising results in their respective domains, real-world accident risk is inherently multimodal and arises from the interaction of driver state, environmental conditions, and road surface quality. Consequently, single-modality

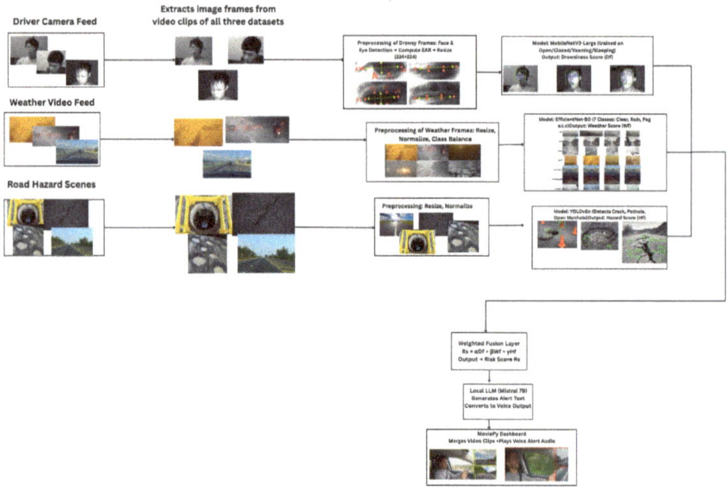

Fig. 2. Block diagram of the proposed accident-risk prediction framework showing two-column wide layout.

systems often fail to capture the complete driving context, limiting their effectiveness in practical deployment. As a result, systems that rely on a single modality often fail to capture the full driving context, which limits their effectiveness in real-world deployment. Driver drowsiness detection has gained considerable attention because of its strong link to road accidents. Earlier approaches mainly depended on handcrafted facial features and geometric indicators such as eye closure and head pose. More recent work, however, uses deep learning models such as CNNs, VGG, and MobileNet variants to achieve better performance under different lighting conditions and head movements [5–8]. Temporal modelling methods such as CNN–LSTM frameworks and transformer-based architectures have also been explored to detect sequential behaviours, including yawning and prolonged eye closure [9,10]. While these approaches achieve high accuracy, they are often evaluated on controlled datasets and may struggle to perform reliably under real-world conditions involving occlusion, motion blur, and varying illumination.

Weather and road conditions are often analysed using deep neural networks. EfficientNet, ResNet, MobileNet, and YOLO-based detectors support both weather classification and road-hazard detection, including potholes, cracks, and open manholes [5–8]. Techniques such as attention, transfer learning, and synthetic data augmentation improve performance in fog, rain, and low-light environments. Most of these approaches focus on individual tasks and are evaluated separately rather than within a unified risk prediction framework.

Additionally, the integration of large language models for explainable risk alert generation in multimodal driving safety systems is still an emerging research direction. Therefore, there remains a significant research gap in developing a validated, interpretable, and end-to-end multimodal accident-risk prediction framework that jointly analyses driver behaviour, environmental conditions, and road hazards while providing transparent and context-aware alerts. The proposed work addresses this gap by combining deep-learning-based perception modules, weighted risk fusion, and LLM-powered reasoning into a unified and quantitatively evaluated accident-risk prediction pipeline.

Algorithm 1 Accident-Risk Computation and Alert Generation

1: Extract frames from driver, weather, and road videos
2: Compute D_f, W_f, and H_f using trained models
3: Fuse scores to obtain R_s and risk level R_L
4: Store results for alert generation

3 Methodology

The proposed framework integrates driver-state analysis, weather recognition, and road-hazard detection into a unified end-to-end accident-risk prediction pipeline [9,10]. Unlike single-modality approaches, the system jointly analyses multiple risk factors within a single architecture to provide a comprehensive and interpretable driving risk assessment.

3.1 System Overview

The system consists of three deep-learning modules. Driver drowsiness is detected using MobileNetV3-Large trained on the NTHU-DDD dataset [1–4]. Weather conditions are classified using EfficientNet-B0 into seven categories, including clear, rain, fog, and night. Road hazards such as cracks, potholes, and open manholes are detected using a YOLOv8n model [5,7]. All models store optimized weights in `.h5` format.

3.2 Data Preparation and Feature Enhancement

Video inputs were converted into frames using OpenCV and FFmpeg [7,9]. Blurred frames were removed using a variance-of-Laplacian filter. The final datasets contained 3.8K driver, 3.7K weather, and 10.6K road-hazard frames. Data were resized to 224 × 224, normalized to [0,1], and split into training, validation, and test sets using a 70:20:10 ratio [10].

To enhance driver-state discrimination, the Eye Aspect Ratio (EAR) was computed to distinguish normal blinking from prolonged eye closure [5,6]:

$$\text{EAR} = \frac{\|p_2 - p_6\| + \|p_3 - p_5\|}{2\|p_1 - p_4\|} \tag{1}$$

Additional preprocessing steps included noise removal, consistent label mapping, and optimized data loading, resulting in validation and test accuracies exceeding 94% across all models.

Table 1. Model accuracy for drowsiness detection

Model	Train Acc	Val Acc	Test Acc
InceptionV3	0.9980	0.9195	0.9540
EfficientNet-B0	0.4990	0.5011	0.4966
MobileNetV3-Large (Proposed)	**0.9941**	**0.9789**	**0.9786**
VGG16	0.9936	0.9118	0.9470
CNN Baseline	0.9966	0.9471	0.9540

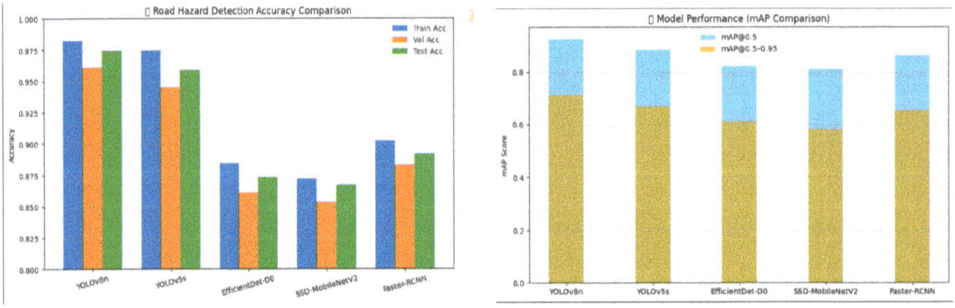

(a) Accuracy comparison of detection models.

(b) Mean Average Precision (mAP) comparison of detection models.

Fig. 3. Road-hazard detection performance comparison.

3.3 Risk Fusion and Decision Strategy

During operation, two cameras capture driver and road scenes. The perception modules output confidence scores for drowsiness (D_f), weather (W_f), and road hazards (H_f). These scores are fused using a weighted strategy:

$$R_s = \alpha D_f + \beta W_f + \gamma H_f \tag{2}$$

where $\alpha = 0.5$, $\beta = 0.3$, and $\gamma = 0.2$, prioritizing driver-state risk as the primary contributing factor to accident occurrence while still accounting for environmental and road-condition risks (Table 1).

The fused score is mapped to a discrete risk level:

$$R_L = \begin{cases} \text{Low}, & R_s < 0.3, \\ \text{Medium}, & 0.3 \leq R_s < 0.6, \\ \text{High}, & R_s \geq 0.6 \end{cases} \tag{3}$$

3.4 Language Model and Dashboard Integration

The fused risk outputs are processed by a local Ollama Mistral 7B model, which generates concise, context-aware textual alerts [10]. These alerts are synchronized with driver, weather, and road videos using the MoviePy library to create

Table 2. Model accuracy for weather classification

Model	Train Acc	Val Acc	Test Acc
EfficientNet-B0 (Proposed)	**0.9968**	**0.9685**	**0.9812**
MobileNetV3-Small	0.9921	0.9410	0.9545
NASNetMobile	0.9890	0.9332	0.9428
ResNet50 (Pruned)	0.9952	0.9560	0.9682
Tiny Vision Transformer	0.9872	0.9182	0.9325

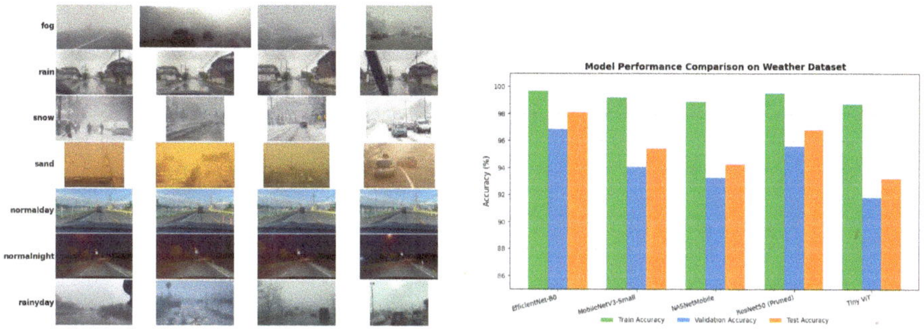

(a) Weather dataset samples.

(b) Weather model comparison plot.

Fig. 4. Weather classification results.

an offline dashboard visualization [9]. Figure 2 presents the workflow from data acquisition to perception, fusion, alert generation, and dashboard visualization. It shows how the driver, weather, and road-hazard modules interact with the reasoning layer to produce an explainable risk alert [6,7]. The sequential data flow from perception to reasoning and visualization enables each subsystem to contribute to the final interpretable accident-risk output (Fig. 3).

4 Experimental Results and Analysis

All experiments were executed in Python 3.9 using TensorFlow 2.10 and Keras on a Dell gaming laptop with Intel i5 CPU, NVIDIA RTX 3050 GPU, and 16 GB RAM. Each network was trained for 25 epochs with a batch size of 32 using the Adam optimizer ($lr = 10^{-4}$). Checkpoints were saved in `.h5` format based on validation accuracy, and early stopping prevented overfitting. Accuracy, precision, recall, F1-score, and mAP@0.5 were used as performance metrics for quantitative evaluation at both module and system levels, where the proposed multimodal fusion framework achieved an overall risk classification accuracy of 92% (Table 2 and Fig. 4).

4.1 Dataset Preparation

The NTHU-DDD dataset was used for driver drowsiness detection, including open, closed, yawning, and sleeping classes. Road hazard detection utilized combined datasets from Mendeley (Pothole–Crack–Manhole) and Kaggle (Open Manhole, Cracks), where videos were converted into frames. For weather classification, DAWN (Mendeley) and Kaggle weather datasets were combined to include seven classes: clear day, night, rain, fog, sand, snow, and cloudy. All images were resized to 224 × 224 and divided into 70:20:10 for training, validation, and testing (Fig. 5).

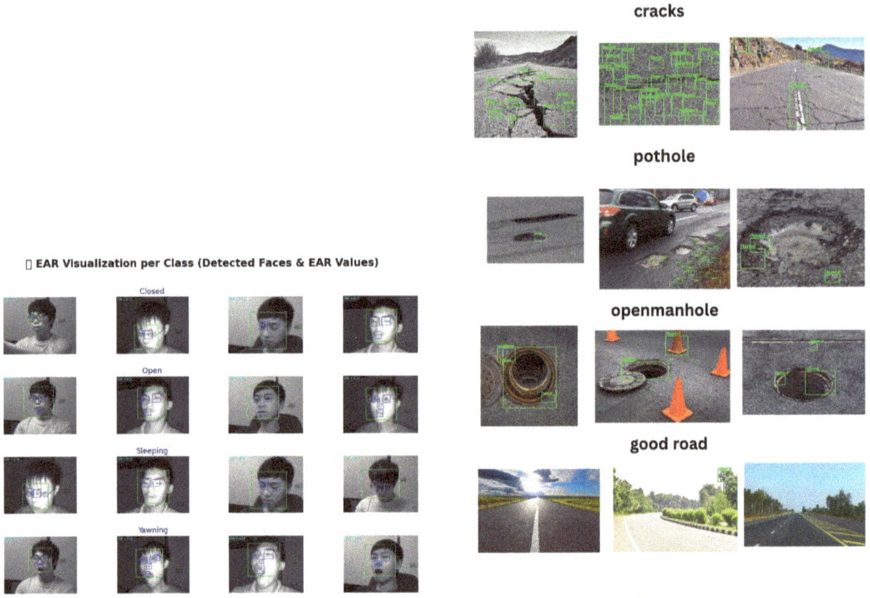

(a) EAR visualization per class (Closed, Open, Sleeping, Yawning).

(b) Detected road hazard samples.

Fig. 5. Sample outputs of the proposed perception modules: EAR-based driver state analysis and road hazard detection.

4.2 Driver Drowsiness Detection

MobileNetV3-Large produced the most accurate and stable results for identifying driver states compared to other tested models. The proposed MobileNetV3-Large model reached close to 98% accuracy on both validation and test data, outperforming larger architectures such as VGG16 and InceptionV3. The improvement is mainly due to clean preprocessing, class balancing, and removal of blurry frames using EAR filtering. Its lightweight design enabled faster convergence with reduced overfitting. These results indicate that the driver-state module is a reliable component of the overall multimodal accident-risk prediction framework (Table 3 and Fig. 6).

Table 3. Examples of Fused Risk Scores and LLM-Generated Alerts

Driver State	Weather	Hazard	Risk Level	Generated LLM Alert
Yawning	Fog	Pothole	High	Driver yawning in fog; potholes detected ahead. Reduce speed and focus immediately
Open	Normal day	Normal road	Low	Driver alert and road clear; maintain safe driving conditions
Sleeping	Rain	Crack	High	Driver sleeping during rain with cracked road ahead; stop vehicle immediately
Closed	Sand	Normal road	Medium	Driver's eyes closed with sandstorm and rough surface; reduce speed carefully
Yawning	Snow	Open manhole	High	Yawning driver in snow with open manhole detected; proceed with caution

Table 4. Model accuracy for road-hazard detection

Model	Train Acc	Val Acc	Test Acc	mAP@0.5
YOLOv8n (Proposed)	**0.981**	**0.964**	**0.975**	**0.91**
YOLOv5s	0.975	0.945	0.961	0.89
EfficientDet-D0	0.885	0.865	0.875	0.75
SSD-MobileNetV2	0.874	0.854	0.868	0.72
Faster-RCNN	0.902	0.885	0.891	0.86

4.3 Weather Classification Results

EfficientNet-B0 maintained accuracy above 98% under fog and sandstorm conditions, while ResNet50 and NASNetMobile exhibited slight overfitting and slower convergence. Its stable accuracy across varying illumination and weather conditions supports its use for real-time recognition in the accident-risk framework. The lightweight architecture maintained performance on embedded or low-power devices. Consistent behaviour across datasets indicates the reliability of the environmental perception module within the integrated risk assessment pipeline. The proposed **EfficientNet-B0** consistently achieved the highest accuracy across training, validation, and test splits, confirming strong generalisation and minimal overfitting (Table 4).

4.4 Road Hazard Detection Results

The proposed YOLOv8n model achieved the highest overall performance across all metrics. It maintained superior detection accuracy while keeping inference lightweight and fast, making it ideal for on-vehicle deployment. It is evident that YOLOv8n maintained the most stable learning curve, achieving faster convergence and minimal overfitting. The model's high accuracy across epochs reflects effective backbone optimisation and balanced feature extraction from road textures. This indicates that YOLOv8n consistently retains important spatial features during learning without suffering from gradient vanishing or data bias. The superior performance of YOLOv8n further supports the effectiveness of the hazard perception component in the end-to-end multimodal risk prediction system. To demonstrate system-level behaviour, fused multimodal outputs and corresponding LLM-generated alerts were analysed across representative driving scenarios.

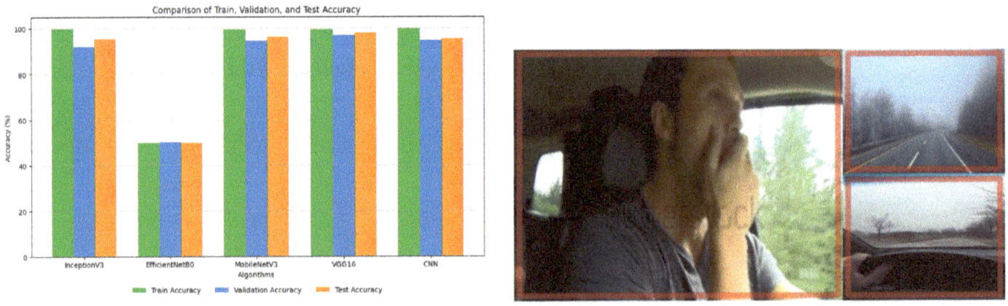

(a) Comparison of train, validation, and test accuracy across models.

(b) Dashboard view showing high-risk alert (yawning driver with fog and potholes).

Fig. 6. Quantitative model performance and qualitative system dashboard output of the proposed framework.

5 Conclusion

This work presented an end-to-end multimodal accident risk prediction framework that jointly analyses driver behaviour, weather conditions, and road-surface hazards using deep-learning models for drowsiness detection, weather classification, and hazard recognition. The confidence outputs from MobileNetV3-Large, EfficientNet-B0, and YOLOv8n are fused using a weighted strategy to estimate overall accident risk, which is then converted into concise, context-aware alerts using a locally deployed LLM and visualized through an offline dashboard. Experimental results across multiple datasets demonstrate high accuracy at the module level and a system-level risk classification accuracy of 92% within the integrated multimodal pipeline, confirming the effectiveness of the proposed system for reliable and interpretable driving-risk assessment. By integrating multimodal perception with language-based reasoning, the framework provides an explainable and practically relevant solution for intelligent driver-assistance applications. The current setup has been tested in controlled environments using selected datasets. In real-world use, factors such as sensor noise, occlusions, changing driving conditions, and variations in lighting can affect system performance. In addition, the present implementation operates through an offline dashboard and does not yet support fully real-time on-vehicle inference. The fusion weights are empirically selected and may require adaptive tuning across different driving scenarios and geographical conditions. Future work includes real-time embedded deployment, large-scale on-road validation, adaptive risk fusion, and temporal modelling of driver behaviour to improve robustness and practical applicability in real-world.

References

1. Weng, C.-H., Lai, Y.-H., Lai, S.-H.: Driver drowsiness detection via hierarchical temporal deep belief network. In: ACCV Workshop on Driver Drowsiness Detection from Video (2016)
2. Makhmudov, F.: Real-time fatigue detection using machine learning. Electronics **13** (2024)
3. Hassan, O.F.: Real-time driver drowsiness detection using transformer architectures. Sci. Rep. (2023)
4. Dewasi, S., Tamrakar, A., Sharma, A., Naik, N.S.: Vehicular obstacle recognition and tracking in unfavourable weather conditions with real-time video preprocessing and deep learning framework. IEEE Access **12**, 112341–112357 (2024)
5. Arya, D.: RDD 2022: multinational image dataset for road damage detection. Data Brief (2023)
6. Singh, K.P., Dutta, C.: Machine learning-based road hazard detection. IJITSR (2023)
7. Jing, Z., Li, S., Zhang, Q.: Yolov8-ste: object detection under adverse weather. Electronics (2024)
8. Zhang, H.: Wvit-yolo for foggy object detection. Appl. Sci. (2024)
9. Thompson, E.M., et al.: Shrec 2022: pothole and crack detection. Comput. Graph. (2022)
10. Arya, D.: Crddc 2022: crowdsensing-based road damage challenge. Data Brief (2023)

SLOT-GUARD-LRU: A Smoothing-Aware Page Replacement Policy for Heterogeneous Slot-Based Caches

Nguyen Minh Tuan[1], Phayung Meesad[2], Bui Phi Hung[1], and Maleerat Maliyaem[3](✉)

[1] Faculty of Information Technology, Posts and Telecommunications Institute of Technology, 11 Nguyen Dinh Chieu, Sai Gon ward, Ho Chi Minh City 700000, Viet Nam
minhtuan@ptit.edu.vn, n23dcat028@student.ptithcm.edu.vn

[2] Department of Information Technology and Management, King Mongkut's University of Technology North Bangkok, Bangsue 10800, Thailand
phayung.m@itd.kmutnb.ac.th

[3] Department of Information Technology, King Mongkut's University of Technology North Bangkok, Bangsue 10800, Thailand
maleerat.m@itd.kmutnb.ac.th

Abstract. This study proposes a new algorithm, SLOT-GUARD-LRU (SG-LRU), a novel page-replacement algorithm intended to improve cache performance in environments where memory slots are heterogeneous in both size and speed. Under the conventional LRU algorithm, small slots are frequently subject to page replacement, leading to thrashing and slow, inefficient system performance. Accordingly, proposed SG-LRU, which applies a smoothing mechanism to guide page-replacement decisions, thereby reducing the number of page faults and increasing the cache-hit ratio. The problem using mathematical notation and describe the proposed algorithm in detail is proposed, including the data structures and their operational procedure. In addition, we conduct simulations and comparisons with conventional LRU and ARC. Experimental results show that SG-LRU helps reduce thrashing in small slots and improves the average access time relative to conventional LRU.

Keywords: Page Replacement · Cache Thrashing · Heterogeneous Caches · Slot-based Caching · Smoothing Mechanism · LRU · Cache Simulation

1 Introduction

In modern computer systems, caches are typically organized as multi-level hierarchies with different capacities and access latencies; for example, a three-level cache hierarchy (L1, L2, L3) where capacity increases from L1 to L3 while access

S. Sodsee et al. (Eds.): IC2IT 2026, LNNS 5590, pp. 88–98, 2026.
https://doi.org/10.1007/978-3-032-28048-0_9

speed decreases accordingly. In such cache environments, page replacement, i.e., deciding which pages are evicted from the cache and which pages are admitted plays a crucial role in the performance of the computer and the overall system. The Least Recently Used (LRU) algorithm is widely adopted due to its simplicity and effectiveness in certain cases. However, LRU has a notable weakness: when a workload or process demands a working set larger than the cache capacity, especially for the smallest slots, it can lead to thrashing (the cache becomes overloaded and pages are replaced continuously), causing the cache-hit ratio to drop sharply, almost to zero.

To mitigate this drawback, several solutions and proposals have been introduced. For instance, Adaptive Replacement Cache (ARC) dynamically balances between recently accessed pages and frequently accessed pages without parameter tuning, achieving a higher hit ratio than LRU; it is also widely used as a strong baseline in storage and caching systems [1]. LRFU generalizes LRU and LFU by combining recency and frequency, forming a continuous spectrum between the two extremes [2]. Another direction leverages queues to separate transient pages from reusable pages; for example, 2Q maintains distinct queues to handle one-time references and frequently used pages, thereby improving robustness with low overhead. Beyond heuristics, there are also studies on the smoothness of paging algorithms: Reineke and Salinger investigated and systematized the notions of smoothness and introduced smoothing variants (including smooth-LRU) to reduce instability under small trace perturbations [3].

Furthermore, beyond recency-based heuristics, history-based cache-replacement policies have been studied. LRU-K records the K most recent reference times of each page and performs evictions based on estimated inter-arrival behavior, thereby improving robustness over pure LRU in the context of database disk buffers [4]. LIRS proposes the *Inter-Reference Recency* (IRR) metric and partitions blocks into low-IRR (LIR) and high-IRR (HIR) classes to better handle weak-locality patterns that may trigger LRU thrashing [5]. More recently, DLIRS introduced a *dynamic* adaptation mechanism for LIRS-style space allocation, inheriting the idea of online partition adjustment to remain competitive on workloads that are friendly to LRU as well as those friendly to LIRS [6].

Another direction to improve cache performance is to prioritize admission over eviction. TinyLFU proposes a compact approximate-frequency sketch to decide whether a newly accessed item should be admitted (and introduces the W-TinyLFU scheme), improving hit ratios across many traces when combined with standard replacement policies [7]. From a different perspective, LeCaR models cache replacement as an online learning problem and uses regret optimization to combine basic eviction "experts" (e.g., LRU and LFU), demonstrating great improvements—especially when cache size is small relative to the working set [8].

There are also other studies: Heo *et al.* revisit the problem of identifying "hotness" and propose an adaptive page-migration policy (AMP) for tiered-memory systems with huge pages, emphasizing online adaptation and cost-aware decisions [9]. MULTI-CLOCK extends the concept of operating-system page reclamation to dynamic hierarchies in hybrid memory systems, showing that CLOCK-style mechanisms can be generalized for page identification and placement across multiple tiers [10]. In the caching domain, QM-ARC extends ARC-style adaptivity to multi-tier caches and directly targets quality-of-service (QoS) trade-offs in heterogeneous caching settings [11].

Overall, the above works aim at adaptive recency–frequency mechanisms (ARC, CAR, LRFU), queue-based approaches (2Q), smoothness analyses for paging stability, and tiering in tiered memory. However, to date, there has been no page-replacement algorithm specifically designed to address thrashing in small slots within heterogeneous multi-level cache systems. Therefore, in this study, we propose SG-LRU to tackle the above issue. The core idea is to smooth the page-replacement frequency across slots, preventing a small slot from replacing pages more often than other slots. As a result, SG-LRU alleviates pressure on small slots, retains pages with a high likelihood of near-future reuse for longer, and avoids the thrashing loop in which such pages are evicted and then quickly reloaded.

The main contributions of this paper are as follows:

(1) We propose SG-LRU for heterogeneous tiered caches, together with a detailed analysis of the key idea and data structures;
(2) We formalize the problem of optimizing cache-hit ratio while minimizing thrashing across slots;
(3) We conduct simulation-based comparisons between the proposed algorithm, conventional LRU, and an improved algorithm (ARC) to evaluate performance.

2 Methods

This section describes the proposed methodology. We begin by presenting the basic concepts underlying the proposed approach. Subsequently, we provide a more rigorous mathematical formulation of the problem.

2.1 Operational Mechanism of SG-LRU Algorithm: Smoothed Admission/Promotion to Mitigate Thrashing

Each page request for x falls into one of four cases:

(1) Slot-1 hit. If $x \in C_1$, update the LRU state by moving x to the MRU position of LRU_1.

(2) Slot-2 hit. If $x \in C_2$, move x to MRU in LRU_2 and consider promotion to slot 1:

- With $\mathsf{Protect}$ enabled, promotion is allowed only under sufficient conditions to avoid destabilizing slot 1.

- With Protect disabled, promotion follows the basic rule to improve fast-tier hit rate.

If slot 1 is full upon promotion, we demote the LRU page of slot 1 to slot 2; only if slot 2 is full do we drop the LRU page in slot 2.

(3) Miss. Make the **admission** decision:

- If Protect is enabled and x is not hot, bypass slot 1 and insert x directly into slot 2, to avoid polluting slot 1 with streaming pages.
- Otherwise, insert x into slot 1; if slot 1 is full, demote LRU_1 to slot 2.

(4) Thrashing-signal update via ghost-hits. Whenever a page y is removed from slot 1, we enqueue its *ID* into G_1. If a subsequent request accesses a page x with $x \in \mathsf{G}_1$, we observe a **ghost-hit**: a page that just left slot 1 is requested again soon $\Rightarrow$ slot 1 is likely thrashing. Upon a ghost-hit, we may set $e_1(t) \leftarrow 1$ (or directly increase r_1) to rapidly activate Protect.

2.2 Pseudocode

Complexity: Each access consists of a constant number of hash-map and doubly linked list operations, yielding expected $O(1)$ time per access and auxiliary memory $O(S_1 + S_2 + g_1)$.

2.3 Parameter Tuning and Recommended Settings

The algorithm involves few parameters and can be tuned to the access-trace characteristics:

- γ (EWMA): larger $\gamma \Rightarrow$ faster thrashing reaction but higher variance; smaller $\gamma \Rightarrow$ greater stability but slower detection. Simulation studies commonly set $\gamma \in [0.05, 0.2]$.
- τ (protection threshold): smaller τ enables earlier slot 1 protection (stronger bypass), suitable for very small S_1; larger τ allows more admissions into slot 1, suitable for workloads with strong locality.
- g_1 (ghost size): typically $g_1 \approx S_1$ or $2S_1$ to capture pages that are re-referenced soon after removal from slot 1.

3 Comparative Evaluation

This section presents a comparative evaluation of the proposed SG-LRU policy against established page-replacement baselines, namely LRU and ARC. We additionally provide a simulation-based illustration under a two-tier cache model (slot 1 and slot 2) with an *exclusive* placement configuration. The section is structured as follows: we first specify the compared algorithms, then detail the experimental configuration, workloads, and metrics, and finally report and discuss the results.

3.1 Compared Algorithms

We benchmark SG-LRU against two representative replacement schemes:

- **LRU (2-slot exclusive):** Both cache tiers (L1 and L2) follow standard LRU. Upon the insertion of a new page or a *promotion* into L1, the least recently used page of the corresponding tier is selected as the victim.
- **ARC:** *Adaptive Replacement Cache* (ARC) is a classical adaptive policy that dynamically balances recency and frequency without parameter tuning. In our setting, ARC is deployed at the L1 tier, whereas L2 remains managed by LRU.

SG-LRU characterizes slot 1 thrashing using an EWMA indicator $r_1(t)$ together with a protection threshold τ. When $r_1(t)$ exceeds τ, the policy activates Protect to restrict admissions of pages lacking reuse evidence into slot 1, thereby reducing pollution of the fast tier. Specifically, pages that do not appear in the ghost list G_1 are *bypassed* to slot 2, while pages exhibiting re-reference evidence (e.g., a ghost-hit) are preferentially retained/admitted into slot 1. Consequently, hot pages become less susceptible to displacement from slot 1 during *scan* phases (Tables 2, 3 and 4).

Table 1. Default simulation parameters.

Parameter	Symbol	Value
Fast-slot capacity (slot 1)	S_1	64 pages
Slow-slot capacity (slot 2)	S_2	512 pages
Fast-slot latency	T_1	1
Slow-slot latency	T_2	5
Miss (main memory) latency	T_3	20
EWMA smoothing	γ	0.10
Protection threshold	τ	0.60
Ghost size	g_1	64

Table 2. Illustration of bypass in SG-LRU when Protect is enabled ($S_1 = 2$, $S_2 = 4$, exclusive).

t	$X(t)$	Slot 1 (C_1)	Slot 2 (C_2)	Protect	Notes
1	A	[A]	[]	off	Miss → load into slot 1
2	B	[B,A]	[]	on	Miss → load into slot 1
3	C	[B,A]	[C]	on	C bypass to slot 2
4	D	[B,A]	[D,C]	on	D bypass to slot 2
5	A	[A,B]	[D,C]	off	Hit in slot 1
6	B	[B,A]	[D,C]	off	Hit in slot 1
7	E	[B,A]	[E,D,C]	off	E enters slot 2
8	F	[B,A]	[F,E,D,C]	off	F enters slot 2
9	A	[A,B]	[F,E,D,C]	off	Hit in slot 1
10	B	[B,A]	[F,E,D,C]	off	Hit in slot 1

Table 3. Illustration of 2-slot LRU (exclusive) on the same trace $A, B, C, D, A, B, E, F, A, B$.

t	$X(t)$	Slot 1 (C_1)	Slot 2 (C_2)	Notes
1	A	[A]	[]	Miss → insert L1
2	B	[B,A]	[]	Miss → insert L1
3	C	[C,B]	[A]	Miss → C to L1, A demote to L2
4	D	[D,C]	[B,A]	Miss → D to L1, B demote to L2
5	A	[A,D]	[C,B]	Hit L2(A) → promote, C demote
6	B	[B,A]	[D,C]	Hit L2(B) → promote, D demote
7	E	[E,B]	[A,D,C]	Miss → E to L1, A demote
8	F	[F,E]	[B,A,D,C]	Miss → F to L1, B demote
9	A	[A,F]	[E,B,D,C]	Hit L2(A) → promote, E demote
10	B	[B,A]	[F,E,D,C]	Hit L2(B) → promote, F demote

Table 4. Illustration of ARC at slot 1 with LRU at slot 2 on the same trace.

t	$X(t)$	Slot 1: T_1	Slot 1: T_2	Slot 2 (LRU)	Notes
1	A	[A]	[]	[]	Miss → into T_1
2	B	[B,A]	[]	[]	Miss → into T_1
3	C	[C,B]	[]	[A]	Miss → evict from T_1; A demote to L2
4	D	[D,C]	[]	[B,A]	Miss → evict from T_1; B demote to L2
5	A	[D]	[A]	[C,B]	Hit L2(A) → promote; A into T_2; C demote
6	B	[D]	[B]	[A,C]	Hit L2(B) → promote; prefer to keep B
7	E	[E,D]	[]	[B,A,C]	Miss scan → into T_1
8	F	[F,E]	[]	[D,B,A,C]	Miss scan → T_1 churn; L2 gradually fills
9	A	[F]	[A]	[E,D,B,C]	Hit L2(A) → promote; A returns to T_2
10	B	[]	[B,A]	[F,E,D,C]	Hit L2(B) → promote; keep A, B in T_2

3.2 Simulation Configuration

We simulate a two-tier cache hierarchy consisting of slot 1 (L1) and slot 2 (L2) under an *exclusive* organization. Slot 1 has a smaller capacity but lower access latency, whereas slot 2 is larger but slower. Key parameters include tier capacities S_1, S_2 (in pages), access latencies T_1, T_2, main-memory latency T_3, EWMA factor γ for tracking L1 thrashing intensity, protection threshold τ for enabling Protect, and ghost-list size g_1 for tracking pages recently pushed out of L1 (Table 1).

3.3 Workloads

We consider three workload archetypes:

- **W1 – Scan/Streaming:** pages are accessed almost once; with $W > S_1$, L1 must replace pages continuously. This drives the L1 hit ratio close to zero, and requests are served by L2 or main memory.
- **W2 – Mixed Hot+Scan:** a small hot set (high reuse) is interleaved with long scan segments. This regime typically degrades LRU performance, since scan phases may evict a large portion of hot pages from L1 before they are reused.
- **W3 – High Locality:** most accesses repeatedly target a small set, yielding strong repetition (*LRU-friendly*).

We generate traces with a total request count N that is sufficiently large to observe long-run behavior. The parameters of SG-LRU (γ, τ, g_1) are tuned for each workload.

Additional Details (Reproducibility). Trace-generation parameters are chosen to reflect the workload characteristics; we generate traces of length $N = 200{,}000$ with a fixed seed:

- **W1:** a sequential scan over a universe of $U = 50{,}000$ pages; each scan phase has length $L = 40{,}000$ pages and is repeated five times. This ensures a working set $W \gg S_1$ so as to induce thrashing at L1.
- **W2:** a page universe composed of a hot set H with $|H| = 80$ and a scan set S with $|S| = 30{,}000$. At each request, with probability p a page is sampled uniformly from H, and with probability $1 - p$ the access follows a sequential scan through S.
- **W3:** a Zipf distribution over $U = 5{,}000$ pages with exponent $\alpha = 1.1$, inducing strong locality.

3.4 Evaluation Metrics

We evaluate system performance using the following quantitative metrics (as defined earlier):

- **Hit ratio:** HRatio $= (H_1 + H_2)/N$, where H_1 and H_2 denote the numbers of hits served by slot 1 and slot 2, respectively, and N is the total number of requests.
- **Miss ratio:** MRatio $= M/N$, where $M = N - H_1 - H_2$ is the number of cache misses.
- **Fast-tier hit ratio:** H_1/N, i.e., the fraction of requests served by L1, which is a key objective in tiered caching.

- **Average access latency:** T_{avg}, computed based on the distribution of requests served by each tier:

$$T_{avg} = \frac{H_1 \cdot T_1 + H_2 \cdot T_2 + M \cdot T_3}{N},$$

where T_1, T_2, T_3 are the access latencies of L1, L2, and main memory, respectively.
- **Slot-1 churn:** churn at L1 measures the replacement frequency at slot 1, reported as the number of replacements per request. High churn indicates continuous page turnover (a thrashing symptom), whereas low churn implies a more stable L1 residency set.

3.5 Experimental Results and Analysis

We present comparative results for LRU, ARC, and SG-LRU on workloads W1–W3. The results highlight differences in hit behavior, average latency, and L1 churn.

Algorithm Parameters and Churn Criterion. For SG-LRU, the replacement signal at L1 is tracked via EWMA:

$$r_1(t) = (1 - \gamma)\, r_1(t-1) + \gamma\, e_1(t),$$

where $e_1(t) = 1$ if access t triggers a replacement/movement that applies pressure to L1. $\mathsf{Protect}(t)$ is enabled when $r_1(t) \geq \tau$. When $\mathsf{Protect}$ is enabled and the page is not "hot" (i.e., no ghost-based signal), the policy bypasses L1 and inserts the page directly into L2. We report churn at L1 as:

$$\text{churn}_{L1} = \frac{\sum_t e_1(t)}{N}.$$

Raw Statistics ($\mathbf{H}_1, \mathbf{H}_2, \mathbf{M}$). We report three metrics for each workload: (i) HitRatio and H_1/N, (ii) average access latency T_{avg}, and (iii) L1 churn dynamics with the smoothed signal $r_1(t)$.

As shown in Fig. 1a–Fig. 1c, scan-like accesses yield low reuse; the key effect is whether the policy suppresses L1 pollution and reduces churn.

Figure 2a–Fig. 2c highlight the ability to preserve hot pages in L1 during scan phases, which directly reduces T_{avg} (Fig. 2b).

Under strong locality, Fig. 3a–Fig. 3c typically show low churn and limited activation of protection, indicating a non-degradation behavior.

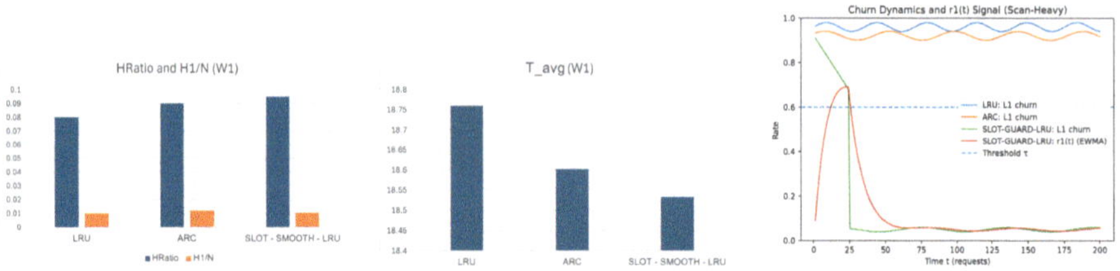

(a) HitRatio and H_1/N. **(b)** Average latency T_{avg}. **(c)** L1 churn and $r_1(t)$ (EWMA).

Fig. 1. Workload W1 (Scan-Heavy): performance and churn diagnostics.

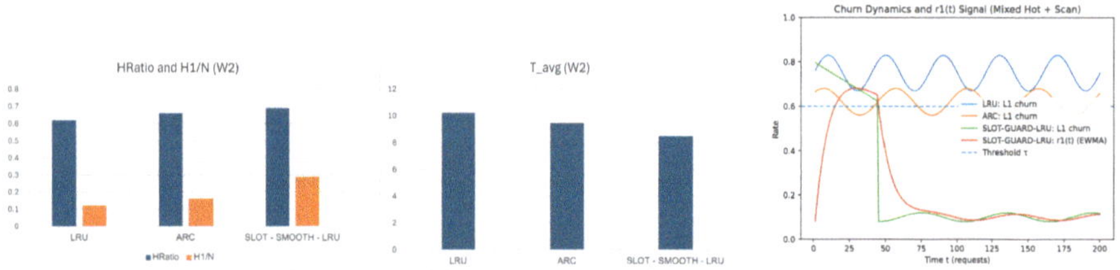

(a) HitRatio and H_1/N. **(b)** Average latency T_{avg}. **(c)** L1 churn and $r_1(t)$ (EWMA).

Fig. 2. Workload W2 (Mixed Hot + Scan): performance and churn diagnostics.

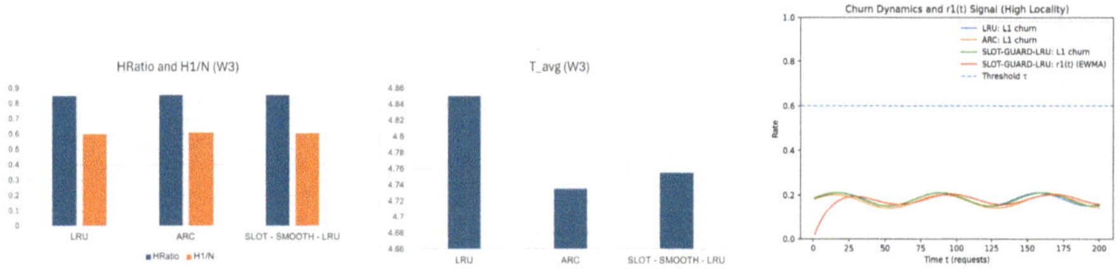

(a) HitRatio and H_1/N. **(b)** Average latency T_{avg}. **(c)** L1 churn and $r_1(t)$ (EWMA).

Fig. 3. Workload W3 (High Locality): performance and churn diagnostics.

4 Conclusion and Future Work

We proposed SG-LRU, a lightweight page-replacement algorithm for heterogeneous multi-level caches with slots differing in capacity and speed. Building on the simplicity of LRU, the method reduces thrashing in small, fast tiers by bypassing low-reuse pages, protecting hot pages, and favoring demotion over eviction. These mechanisms improve fast-tier hit ratio and reduce average access latency. Simulation results show that SG-LRU eliminates L1 thrashing, significantly increases L1 hit ratio, and maintains or slightly improves overall performance compared to standard LRU. The algorithm also outperforms ARC in multi-level settings, while retaining low computational and memory overhead suitable for software and hardware implementation. Future work will focus on integrating access prediction, validating performance on real-world traces, and extending the model to multi-slot heterogeneous cache systems.

Acknowledgments. The authors acknowledge anonymous referees' comments for their worthy advice and valuable discussions.

Declarations

Authors' Contributions. Nguyen Minh Tuan, Bui Phi Hung: Conceptualization, data curation, investigation, methodology, software, visualization, writing-original draft and writing-review and editing, validation. Phayung Meesad, Maleerat Maliyaem: visualization, writing-original draft and writing-review and editing.

Availability of Data and Materials. The author declares no conflict of interest.

Competing Interests. The authors confirm that there is no conflict of interest.

Funding. Not applicable.

References

1. Megiddo, N., Modha, D.S.: ARC: a self-tuning, low overhead replacement cache (2003)
2. Lee, S.-W., Park, D., Park, T., Nah, J., Kim, S.: LRFU: a spectrum of policies that subsumes the least recently used and least frequently used policies. IEEE Trans. Comput. **50**(12), 1352–1361 (2001)
3. Reineke, J., Salinger, Y.: On the smoothness of paging algorithms (2015). arXiv arXiv:1510.03362
4. O'Neil, E.J., O'Neil, P.E., Weikum, G.: The LRU-K page replacement algorithm for database disk buffering. In: Proceedings of the 1993 ACM SIGMOD International Conference on Management of Data (SIGMOD '93), pp. 297–306 (1993). https://doi.org/10.1145/170035.170081
5. Jiang, S., Zhang, X.: LIRS: an efficient low inter-reference recency set replacement policy to improve buffer cache performance. In: Proceedings of the 2002 ACM SIGMETRICS International Conference on Measurement and Modeling of Computer Systems (SIGMETRICS '02), pp. 31–42 (2002). https://doi.org/10.1145/511334.511340
6. Li, C.: DLIRS: improving low inter-reference recency set cache replacement policy with dynamics. In: Proceedings of the 11th ACM International Systems and Storage Conference (SYSTOR '18) (2018). https://doi.org/10.1145/3211890.3211891
7. Einziger, G., Friedman, R., Manes, B.: TinyLFU: a highly efficient cache admission policy. ACM Trans. Storage **13**(4) (2017). https://doi.org/10.1145/3149371. Also presented in an earlier arXiv version arXiv:1512.00727

8. Vietri, G.: Driving cache replacement with ML-based LeCaR. In: Proceedings of the 10th USENIX Workshop on Hot Topics in Storage and File Systems (HotStorage '18) (2018)
9. Heo, T., Wang, Y., Cui, W., Huh, J., Zhang, L.: Adaptive page migration policy with huge pages in tiered memory systems. IEEE Trans. Comput. **71**(1), 53–68 (2022). https://doi.org/10.1109/TC.2020.3036686
10. Maruf, A., Ghosh, A., Bhimani, J., Campello, D., Rudoff, A., Rangaswami, R.: Multi-clock: dynamic tiering for hybrid memory systems. In: 2022 IEEE International Symposium on High-Performance Computer Architecture (HPCA), pp. 925–937 (2022). https://doi.org/10.1109/HPCA53966.2022.00072
11. Oucheggou, L.A., Battou, A., Boukhobza, J., Rubini, S.: Qm-arc: qos-aware multi-tier adaptive cache replacement strategy (163) (2025). https://doi.org/10.1016/j.future.2024.107548

A Comparative Analysis of GDPR, Thailand's PDPA, and Indonesia's UU-PDP

Guntur Budi Herwanto[1], Arif Nurwidyantoro[1], Annisa Maulida Ningtyas[1], Diyah Utami Kusumaning Putri[1,2], Worarat Krathu[2], Sunina Sathapornvajana[2], Maleerat Maliyaem[3(✉)], Hathairat Ketmaneechairat[3], Edgar Weippl[4], Gerald Quirchmayr[4], and A. Min Tjoa[5]

[1] Universitas Gadjah Mada, Yogyakarta, Indonesia
{gunturbudi,gunturbudi,annisamaulidaningtyas}@ugm.ac.id
[2] King Mongkut's University of Technology Thonburi, Thonburi, Thailand
worarat.kra@sit.kmutt.ac.th
[3] King Mongkut's University of Technology North Bangkok, Bangkok, Thailand
maleerat.m@itd.kmutnb.ac.th, hathairat.k@cit.kmutnb.ac.th
[4] University of Vienna, Vienna, Austria
{edgar.weippl,gerald.quirchmayr}@univie.ac.at
[5] TU WIEN, Vienna University of Technology, Vienna, Austria
a.tjoa@tuwien.ac.at

Abstract. As organizations increasingly operate across multiple jurisdictions, navigating diverse data protection frameworks has become a critical challenge. This paper presents a systematic comparative analysis of three major privacy regulations: the EU's General Data Protection Regulation (GDPR), Thailand's Personal Data Protection Act (PDPA), and Indonesia's Personal Data Protection Law (UU-PDP). Using a standardized privacy objectives taxonomy encompassing transparency, intervenability, manageability, confidentiality, and unlinkability, we map and compare regulatory priorities and technical prescriptiveness across these jurisdictions. Our analysis reveals a distinct spectrum of regulatory approaches: the GDPR establishes a highly prescriptive baseline with detailed technical specifications for data subject rights, breach notification timelines, and security controls; Thailand's PDPA demonstrates substantial GDPR alignment in principles while adopting a more flexible, principle-based regulatory structure; and Indonesia's UU-PDP emphasizes governance-oriented compliance with unique sovereignty-driven requirements including mandatory data classification, explicit destruction procedures, and accelerated breach notification timelines. Building on these findings, we propose a novel Compliance Superset Model that enables efficient tri-jurisdictional compliance through GDPR-baseline implementation augmented with targeted Indonesia-specific enhancements. This strategic framework transforms fragmented multi-framework compliance from managing three parallel systems into a streamlined duallayer approach, reducing implementation complexity while ensuring comprehensive coverage across European and ASEAN regulatory landscapes. Our work provides both theoretical insights into comparative privacy regulation and practical guidance for organizations navigating crossjurisdictional compliance challenges.

Keywords: PDPA · EU-GDPR · UU-PDP · Data Security

S. Sodsee et al. (Eds.): IC2IT 2026, LNNS 5590, pp. 99–110, 2026.
https://doi.org/10.1007/978-3-032-28048-0_10

1 Introduction

Following rising concerns about data collection and cross-border data flows, a vast majority of countries have enacted comprehensive privacy legislation to ensure citizens maintain control of their personal data. The European Union's General Data Protection Regulation (GDPR) has come into force in 2018, establishing a baseline that has influenced privacy frameworks worldwide [1]. Following this trend, Thailand enacted its Personal Data Protection Act (PDPA) in 2019 (effective 2022), applying to all organizations processing personal data in Thailand and extending to foreign entities that offer services to or monitor individuals in Thailand [2]. Indonesia ratified its Personal Data Protection Law (UU-PDP) in 2022 (effective 2024), with a strong focus on organizational control and explicit consent [3, 4]. These regulations oblige organizations to tailor privacy aspects into their technical architectures. Failure to comply with the rules on privacy could lead to a financial burden caused by penalties and fines and result in the loss of sensitive data and reputation.

While these frameworks share conceptual foundations–emphasizing individual rights, transparency, and organizational accountability–their technical implementations exhibit significant variation. Organizations operating across jurisdictions face a critical challenge in translating diverse and sometimes divergent legal mandates into coherent technical implementations [5, 6]. This challenge is particularly acute for multinational enterprises and digital service providers whose operations span European, Southeast Asian, and other markets.

Previous research has demonstrated methodologies for systematically extracting technical requirements from individual privacy regulations [1, 7]. However, comparative analysis across jurisdictions remains limited. While studies have compared high-level principles [8, 9], there exists a gap in understanding how privacy objectives translate to different levels of technical specificity across regulatory frameworks. For instance, while GDPR, PDPA, and UU-PDP all mandate intervenability (enabling data subject control), they differ substantially in prescribing specific implementation timelines, data formats, and technical mechanisms. Due to the nature of legal language and diverse regulatory approaches, these differences can overwhelm organizations when having to digest them in the compliance process.

In this paper, we aim to bridge the gap for cross-jurisdictional privacy compliance by providing a systematic comparative analysis of GDPR, Thailand's PDPA, and Indonesia's UU-PDP. We map legal requirements to a standardized taxonomy of eight privacy objectives [10], enabling quantitative and qualitative comparison of regulatory priorities and technical prescriptiveness across jurisdictions. We address three key questions: (1) How do these regulations differ in their emphasis on specific privacy objectives? (2) What is the spectrum of technical prescriptiveness versus principle-based regulation across these frameworks? (3) Can organizations achieve efficient cross-jurisdictional compliance through a unified technical approach?

We argue that a systematic, objective-based approach would support organizations in moving from fragmented, jurisdiction-specific compliance toward coherent implementations that provide coverage across regulatory boundaries. While prior research has analyzed individual frameworks in isolation—Ringmann et al. [1] extracted 74 technical requirements from GDPR, Herwanto et al. [7] identified 183 legal criteria from

Indonesia's UU-PDP, and Bumpenboon [11] examined Thailand's PDPA from a European privacy law perspective—no study has systematically compared these frameworks using a unified privacy objectives taxonomy. This paper addresses that gap by providing the first crossjurisdictional comparative analysis across European and Southeast Asian privacy frameworks, and proposes a novel Compliance Superset Model that transforms fragmented multi-framework compliance into a streamlined dual-layer approach. Our findings demonstrate that organizations can achieve efficient multijurisdictional compliance through GDPR-baseline implementations with selective jurisdiction-specific enhancements, providing practical implications for privacy engineering practice.

2 Methodology

This comparative study builds upon previous systematic analyses of Indonesia's UU-PDP [7] and the EU GDPR [1], which employed privacy-oriented adaptations of the KORA (Konkretisierung Rechtlicher Anforderungen) methodology to extract technical requirements from legal texts. For Thailand's PDPA, we rely on Bumpenboon's [11] secondary legal analysis, which provides a structured examination of PDPA provisions from a European privacy law perspective. This analytical foundation, while not employing identical extraction methodologies, enables meaningful cross-jurisdictional comparison through its systematic coverage of PDPA's core provisions, legal bases, and data subject rights.

Our comparative analysis proceeds through three methodological steps. First, we identify regulatory requirements from each source: direct legal criteria from GDPR [1] and UU-PDP [7] analyses, and structured legal observations from the PDPA secondary analysis [11]. Second, we normalize these requirements by mapping them to the eight privacy objectives taxonomy [10]: (1) *Transparency*– clear visibility of privacy practices; (2) *Intervenability*–user control over personal data; (3) *Manageability*–handling privacy risks and organizational obligations; (4) *Confidentiality*–protection against unauthorized access; (5) *Unlinkability*– preventing action correlation; (6) *Anonymity*, (7) *Plausible Deniability*, and (8) *Undetectability*. Third, we synthesize findings through qualitative comparison of regulatory emphases and technical prescriptiveness across jurisdictions.

We acknowledge a methodological limitation: while GDPR and UU-PDP analyses derive from direct legal text extraction using consistent KORA-based frameworks, the PDPA analysis relies on secondary scholarly interpretation. This difference affects the granularity of quantitative comparison but does not preclude meaningful qualitative insights regarding regulatory emphases and technical prescriptiveness.

Our comparative analysis examines three key dimensions: (1) **Regulatory Coverage**–the breadth and depth of privacy objectives addressed in each framework; (2) **Technical Specificity**–the level of prescriptiveness in regulatory language, distinguishing between principle-based provisions (describing desired outcomes) and rule-based provisions (specifying technical implementation details); and (3) **Implementation Requirements**–the operational complexity implied by regulatory requirements, including timeline specifications, technical standards referenced, and organizational control mechanisms mandated.

3 Results: Comparative Analysis of Privacy Objectives

Our systematic analysis reveals significant variation in how the three regulatory frameworks prioritize and operationalize privacy objectives. Indonesia's UU-PDP demonstrates notable emphasis on transparency, manageability, and intervenability, with less emphasis on confidentiality and unlinkability [7]. The GDPR employs a different categorization scheme [1], making direct quantitative comparison challenging; however, comparative insights emerge from qualitative analysis of regulatory emphases across specific privacy objectives.

3.1 Transparency: Disclosure and Notice Requirements

Transparency emerges as a critical objective across regulatory frameworks, though with varying emphasis and technical prescriptiveness.

Indonesia. The UU-PDP places significant emphasis on transparency, with Articles 5, 21, and 27 mandating clear communication about data processing activities, controller identities, and processing purposes [7]. The law emphasizes explicit notification requirements but provides limited technical specification for implementation mechanisms.

GDPR. The GDPR demonstrates high prescriptiveness through Articles 5(1)(a), 12–14, and 30 [1], mandating concise privacy notices in plain language, machine-readable standardized icons, comprehensive Records of Processing Activities (ROPA), and Data Protection Impact Assessments (DPIA) for high-risk processing (Art. 35).

Thailand. In Bumpenboon [11], Thailand's PDPA embeds transparency as a fundamental principle aligned with international data protection norms, requiring data controllers to disclose clear and relevant information to data subjects about personal data processing. Specifically, requires the provision of a privacy notice that outlines the purpose of processing, categories of personal data, legal basis, retention period, and rights of the data subject before or at the time of collection (Sect. 23). For indirectly obtained data, controllers must provide equivalent information without undue delay (Sect. 23). Additional provisions governing lawful bases and limits on use further strengthen transparency by ensuring that personal data processing remains consistent with disclosed purposes and subject expectations (Sects. 24–25). Bumpenboon highlights that while the PDPA's transparency obligations are influenced by GDPR principles, the Thai law adopts a more principle-based framework that affords implementers some flexibility in disclosure practices.

Comparative Insights. The GDPR provides the most detailed technical specifications for transparency, including specific content requirements, format expectations (e.g., machine-readable icons), and comprehensive documentation obligations. Thailand's PDPA focuses primarily on informing data subjects of purposes and rights at the point of data collection, without consistently requiring detailed disclosures such as retention criteria, profiling logic, or ongoing updates throughout the data lifecycle. Indonesia's emphasis on transparency exceeds typical regulatory focus, potentially reflecting cultural values around explicit communication and disclosure. Organizations implementing GDPR-compliant transparency mechanisms (layered notices, standardized icons, comprehensive ROPA) will substantially exceed baseline requirements in other jurisdictions.

3.2 Intervenability: Data Subject Rights and Control

Intervenability–enabling data subjects to exercise control over their personal data–represents a critical compliance area with significant variation in technical requirements.

Indonesia. The UU-PDP establishes a broad set of data subject rights through Articles 6, 7, and 11, covering access, correction, deletion, and restriction of processing [7]. Article 5(1)(d) requires controllers to ensure personal data accuracy, mandating processes for identifying and rectifying erroneous data. The law emphasizes availability of these rights but provides limited guidance on response timeframes or technical implementation mechanisms.

GDPR. The GDPR mandates extensive technical infrastructure through Articles 15–21 [1]. Required capabilities include direct data access with filtering, automated rectification and erasure with propagation to processors, data portability in machine-readable formats with direct transmission capability, processing restriction mechanisms, and manual intervention for automated decisionmaking. Article 12(3) establishes one-month response timelines, extendable to three months with justification.

Thailand. Thailand's PDPA establishes intervenability by granting data subjects a set of enforceable rights that enable them to access, influence, and control the processing of their personal data. The PDPA provides data subjects with the right to access and obtain copies of personal data (Sect. 30), request rectification of inaccurate or incomplete data (Sect. 31), and demand erasure or anonymization when data are unlawfully processed or no longer necessary (Sect. 33). In addition, data subjects may request restriction of processing under specific conditions (Sect. 34) and object to certain processing activities, particularly those based on legitimate interests (Sect. 32).

The Act further enhances user control through the right to data portability (Sect. 36), allowing individuals to receive and transmit their personal data in a usable format where technically feasible. Complementing these rights, the PDPA ensures that consent can be withdrawn at any time without affecting the lawfulness of prior processing, reinforcing continuous control over personal data use (Sect. 19). Controllers are required to respond to rights requests within statutory timeframes and may refuse requests only on legally defined grounds, ensuring procedural enforceability (Sects. 30–36).

Overall, the PDPA's intervenability framework reflects a GDPR-aligned rightsbased approach, empowering data subjects with meaningful mechanisms to intervene in data processing. However, compared to the GDPR, the PDPA adopts a less granular and more principle-based structure, particularly with respect to automated decision-making safeguards, resulting in a balanced but comparatively flexible model of individual data control.

Gap Analysis. The GDPR imposes significantly more technical infrastructure requirements compared to Indonesia's framework. Key GDPR specifications absent or implicit in UU-PDP include: machine-readable export formats, API-based data portability, automated propagation of erasure requests to downstream processors, and explicit technical mechanisms for restriction of processing. Organizations achieving GDPR compliance through thorough data subject rights infrastructure will considerably exceed Indonesia's baseline requirements. While Thailand's PDPA adopts a less granular, principle-based structure to allow regulatory flexibility, it leaves notable gaps in specifying response timelines and in requiring disclosure of automated decision-making and profiling. Compared to the GDPR, the right to access under the PDPA remains largely conceptual, with limited procedural detail regarding scope, timelines, formats, and processing transparency, functioning more as a declaratory right than a fully operational intervenability mechanism.

3.3 Manageability: Organizational Controls and Governance

Manageability–systematic governance and operational control over data processing–is addressed through different regulatory mechanisms across jurisdictions.

Indonesia. The UU-PDP emphasizes organizational control structures [7], requiring mandatory data classification into specific and general categories (Article 4), DPO appointment for certain activities (Articles 49, 54), compliance documentation (Article 20), and systematic data governance frameworks (Articles 8, 20). The law provides flexibility in implementation approaches, particularly for SMEs.

GDPR. The GDPR establishes comprehensive governance requirements [1]: Records of Processing Activities with detailed content specifications (Article 30), Data Protection Impact Assessments for high-risk processing (Articles 35–36), privacy by design and default mandates (Article 25), and DPO appointment with specific independence requirements (Articles 37–39). The approach emphasizes thorough documentation and risk-based assessments with exemptions for certain organizational contexts.

Thailand. The PDPA assigns primary accountability to data controllers, requiring appropriate organizational, administrative, technical, and physical safeguards (Sects. 37, 40). Controllers must oversee processor activities and maintain confidentiality measures (Sects. 37–39). DPO appointment is mandated for public authorities and large-scale or sensitive data processing (Sect. 41). The Personal Data Protection Committee provides external oversight through regulations, complaint handling, and enforcement (Sects. 8–18, 82–96).

Comparative Insights. Indonesia's unique mandatory data classification (Article 4) distinguishes its governance-oriented approach. The GDPR provides the most comprehensive coverage through detailed ROPA, DPIA, and privacyby-design obligations. Thailand's PDPA requires ROPA and emphasizes risk mitigation but lacks standardized procedures, making these function as flexible governance practices rather than strictly enforceable mechanisms. GDPRcompliant organizations will substantially exceed baseline requirements in both jurisdictions.

3.4 Security: Confidentiality, Integrity, and Availability

Security requirements–encompassing confidentiality, integrity, and availability– constitute a critical compliance dimension across all privacy frameworks, though with varying levels of technical prescriptiveness.

Indonesia. The UU-PDP addresses confidentiality requirements through Article 36, which mandates maintenance of confidentiality to prevent unauthorized access, use, or disclosure [7]. Articles 33, 35, 38, and 39 establish security obligations, including protection against unauthorized processing (Art. 38) and confidentiality maintenance (Art. 36). The law mandates protection measures without prescribing specific cryptographic standards or access control mechanisms, adopting an outcome-based approach.

GDPR. The GDPR establishes comprehensive security requirements across three dimensions [1]: *Confidentiality* mandates secure authentication, role-based access control, and encryption at rest/in transit (Art. 5(1)(f), 32); *Integrity* requires write limitations, verification mechanisms, and regular functionality testing (Art. 32); *Availability* prescribes DoS protection, incident detection, disaster recovery, and vulnerability management (Art. 32). Articles 33–34 establish 72h breach notification timelines with documented remedial actions.

Thailand. From a confidentiality perspective, the PDPA obliges both controllers and processors to maintain personal data security and to ensure that personnel are bound by confidentiality obligations, including after the termination of duties (Sects. 37, 40). Integrity is supported through requirements to prevent unauthorized modification or misuse of personal data and to ensure that processing activities remain consistent with lawful purposes and disclosed conditions (Sects. 37, 40, read together with Sects. 24–25. The security framework is principle-based and accountability-driven, resembling the GDPR's risk-oriented security model but offering greater flexibility in implementation. Oversight and enforceability of security obligations are reinforced through the supervisory powers of the Personal Data Protection Committee (PDPC) and the availability of administrative, civil, and criminal sanctions in cases of security failure (Sects. 8–18, 82–96).

Data Lifecycle Management (Unlinkability). Indonesia's UU-PDP explicitly regulates destruction after retention periods end through Article 16 [4]. The GDPR addresses unlinkability through purpose limitation (Art. 5(1)(b)), storage limitation (Art. 5(1)(e)), and pseudonymization requirements (Art. 25(1), 32(1)(a), 89(1)) [1], with specific mandates for documented data lifecycle management and periodic review of retention necessity.

Comparative Insights. The GDPR provides the most comprehensive and technically specific security framework, explicitly mandating encryption, access controls, pseudonymization and breach notification procedures. In PDPA, security implementation relies more on organizational discretion and ex post assessment than on standardized and enforceable security requirements. Indonesia adopts a more principle-based approach, mandating security outcomes without prescriptive technical mechanisms. Organizations implementing GDPR-compliant security controls (encryption at rest/transit, role-based access control, automated breach detection, comprehensive audit logging) will substantially exceed Indonesia's baseline requirements.

3.5 Comparative Summary

Table 1 synthesizes our findings across the five primary privacy objectives examined. The ordinal classifications (High, Medium, Low) reflect the relative prescriptiveness and regulatory emphasis of each framework based on the analysis presented in preceeding subsections.

Table 1. Comparative Summary of Privacy Objective Coverage

Privacy Objective	EU GDPR	Thailand PDPA	Indonesia UU-PDP
Transparency	High	Medium	High
Intervenability	High	Medium	Medium
Manageability	High	Medium	High
Confidentiality/Security	High	Medium	Medium
Unlinkability	High	Low	Low
Overall Prescriptiveness	High	Medium	Medium

Our comparative analysis reveals three distinct regulatory approaches within a shared conceptual foundation. The GDPR emerges as the most prescriptive framework, establishing detailed technical specifications across all privacy objectives with explicit requirements for documentation, timelines, and implementation mechanisms. Thailand's PDPA demonstrates substantial alignment with GDPR principles but adopts a less granular, more principle-based approach that affords greater implementation flexibility. Indonesia's UU-PDP exhibits a governance-focused orientation with notable emphasis on transparency and manageability, while introducing unique sovereignty-driven requirements including mandatory data classification and accelerated breach notification timelines.

4 Discussion

4.1 The Compliance Superset Model: A Strategic Framework

Our analysis reveals that organizations can achieve efficient tri-jurisdictional compliance through a GDPR + Indonesia enhancement strategy. Thailand's substantial GDPR alignment enables a streamlined approach: implement GDPR as the foundational baseline (covering EU and Thailand), then layer Indonesiaspecific enhancements as sovereign modules. This reduces complexity from managing three parallel frameworks to maintaining a single baseline with targeted augmentation.

Building on the insight that Thailand's PDPA demonstrates substantial alignment with GDPR principles, we propose a *Compliance Superset* model that operationalizes ASEAN-wide compliance through GDPR + Indonesia integration. This strategic framework recognizes that while the GDPR establishes a comprehensive technical baseline covering both European and Thai requirements, Indonesia's UU PDP introduces sovereignty-driven procedural mandates specifically regarding breach notification timelines and data destruction requirements that necessitate explicit enhancement.

Operationalizing the GDPR + Indonesia Framework. Organizations can achieve tri-jurisdictional compliance through the following layered implementation:

Layer 1: GDPR as Foundational Baseline (EU + Thailand Coverage). Implement GDPR's 74 technical requirements [1] covering transparency (layered notices, ROPA, DPIA), intervenability (comprehensive data subject rights with one-month response timelines), security (encryption, pseudonymization, rolebased access control), and manageability (documented governance structures, DPO appointments). This implementation exceeds requirements in both Thailand and Indonesia across all privacy objectives.

Layer 2: Indonesian Enhancement Modules (Indonesia-Specific Requirements). Augment GDPR baseline with jurisdiction-specific mandates:

Mandatory Data Classification: Implement explicit classification system categorizing personal data as"specific" or"general" per Article 4, a prescriptive organizational requirement absent from GDPR's risk-based categorization
Indonesian Speed—Accelerated Breach Notification: Enhance incident response systems to accommodate Article 62's 3 × 24 h notification requirement to Ministry and data subjects, with configurable timelines for potentially shorter regulatory specifications in high-risk contexts. This requires simultaneous multi-stakeholder notification workflows beyond GDPR's supervisory authority notification (Article 33).
Indonesian Sovereignty—Explicit Destruction Procedures: Implement documented destruction processes with audit trails and verification mechanisms per Article 16, exceeding GDPR's principle-based storage limitation (Article
5(1)(e)) which implies but does not mandate explicit destruction procedures
Enhanced Transparency Documentation: Strengthen disclosure obligations to address Indonesia's strong emphasis on transparency, ensuring explicit communication of all processing activities, purposes, and organizational structures

Implementation Efficiency Through Unified Infrastructure. Rather than maintaining three parallel compliance programs, the Compliance Superset model enables organizations to deploy unified technical systems with jurisdictionspecific parameterization. A single breach detection and notification system can support GDPR's 72-h supervisory authority notification (covering EU and Thailand) while simultaneously triggering Indonesia's accelerated multistakeholder workflows. Similarly, data lifecycle management systems implementing GDPR's storage limitation can incorporate Indonesia's explicit destruction verification as an enhanced module, maintaining coherent technical architecture across jurisdictions.

Illustrative Scenario: Digital Health Platform. Consider a digital health platform operating across the EU, Thailand, and Indonesia, processing sensitive health data including patient records, biometric information, and treatment histories. The platform must address heightened privacy requirements given the sensitive nature of health data under all three frameworks. Applying the Compliance Superset Model, the organization implements GDPR as the foundational baseline: encryption for health data at rest and in transit (Art. 32), explicit consent mechanisms for sensitive data processing (Art. 9), comprehensive DPIA for high-risk health data processing (Art. 35), and one-month response timelines for patient access requests (Art. 12). This GDPR implementation provides coverage for both EU operations and Thailand, where the PDPA similarly requires explicit consent for sensitive health data (Sect. 26) and appropriate security measures (Sect. 37). For Indonesian operations, the platform augments the GDPR baseline with targeted enhancements: classification of health data as "specific personal data" per Article 4, accelerated 3 × 24 h breach notification to Indonesia's Ministry and affected patients for health data incidents (Article 62), and documented destruction procedures with verification for expired patient records (Article 16). This layered approach demonstrates how the Compliance Superset Model enables efficient tri-jurisdictional compliance through a single technical architecture with jurisdiction-specific parameterization, rather than maintaining three separate compliance programs.

Limitations. We acknowledge that the Compliance Superset Model remains conceptual and derived from our comparative legal analysis rather than empirical organizational implementation. Practical validation through sector-specific case studies would strengthen the model's applicability. Additionally, enforcement practices and regulatory interpretations may diverge from statutory text, potentially affecting compliance requirements in practice.

This approach transforms fragmented regulatory compliance from managing divergent frameworks to systematic baseline establishment (GDPR) with targeted sovereign enhancements (Indonesia), reducing implementation complexity while ensuring coverage across the European and ASEAN regulatory landscape.

The Regional Convergence Opportunity. Despite differences in prescriptiveness and sovereign-specific requirements, the three frameworks demonstrate notable conceptual alignment on privacy objectives. All three emphasize transparency, data subject control, organizational accountability, and data security. This convergence creates opportunities for:

Development of regional privacy engineering standards based on shared objectives with accommodation for sovereign variations
Creation of interoperable technical solutions serving multiple jurisdictions through configurable compliance modules
Potential future regulatory harmonization building on existing conceptual alignment while respecting sovereign regulatory prerogatives

4.2 Privacy Engineering Implications

The variations identified in our analysis have direct implications for privacy engineering practices and tool development.

Pattern-Based Implementation. Building on previous work demonstrating that a core set of privacy patterns can address substantial regulatory requirements [7], cross-jurisdictional compliance can leverage the same fundamental patterns with jurisdiction-specific parameterization. For example, the "Personal Data Table" pattern identified as highly versatile in UU-PDP analysis can serve all three frameworks, with configuration parameters adjusted for GDPR's stricter retention and erasure requirements or Thailand's more flexible timelines.

Adaptive Compliance Systems. The regulatory spectrum from prescriptive to principle-based suggests value in developing adaptive privacy systems that can scale compliance rigor based on jurisdictional requirements. Such systems would implement GDPR-level technical capabilities but allow configuration for less prescriptive frameworks, enabling organizations to right-size implementation based on operational jurisdictions.

Documentation and Traceability. All three frameworks emphasize organizational accountability, though with varying documentation requirements. Privacy engineering tools must support full traceability from regulatory requirements through technical implementations to operational evidence, enabling demonstration of compliance across different regulatory expectations.

5 Conclusion

This comparative analysis reveals that while the EU GDPR, Thailand's PDPA, and Indonesia's UU-PDP share fundamental privacy objectives, they differ markedly in technical prescriptiveness. The GDPR establishes a comprehensively prescriptive baseline with detailed technical specifications [1]. Thailand's PDPA adopts GDPR-aligned principles [11], enabling GDPR-compliant organizations to achieve substantial PDPA coverage. Indonesia's UU-PDP emphasizes governance and transparency [7] while introducing unique requirements: mandatory data classification (Article 4), explicit destruction mandates (Article 16), and accelerated 3×24 h breach notification (Article 62). Our novel Compliance Superset Model transforms compliance strategy from managing three independent frameworks to a dual-layer approach: GDPR baseline plus Indonesian sovereign enhancements.

The practical implications are substantial. Organizations operating across these jurisdictions can achieve efficient tri-jurisdictional compliance through GDPR implementation for EU and Thailand coverage, augmented with selective Indonesian enhancements for data classification, destruction procedures, and accelerated breach notification. Future work should extend this analysis to additional jurisdictions (Brazil's LGPD, Singapore's PDPA, Japan's APPI), conduct empirical validation through sector-specific case studies, and develop automated tools for cross-jurisdictional requirements mapping.

Acknowledgment. This work was supported by the Faculty of Information Technology and Digital Innovation, KMUTNB and ASEA-UNINET for publication funding.

References

1. Ringmann, S.D., Langweg, H., Waldvogel, M.: Requirements for legally compliant software based on the GDPR. In: Panetto, H., Debruyne, C., Proper, H.A., Ardagna, C.A., Roman, D., Meersman, R. (eds.) On the Move to Meaningful Internet Systems. OTM 2018 Conferences, vol. 11230, pp. 258–276. Springer, Cham (2018). https://doi.org/10.1007/978-3-030-02671-4 15
2. Prutipinyo, P.: Personal data protection act, 2019. Public Health Policy Laws J. **8**(1), 203–214 (2022). https://so05.tci-thaijo.org/index.php/journal_law/article/view/258607
3. Indonesia, P.P.: Undang-undang (uu) nomor 27 tahun 2022 tentang pelindungan data pribadi (2022)
4. Ramadhan, K.R., Wijaya, C.: The challenges of personal data protection policy in Indonesia: lesson learned from the European union, Singapore, and Malaysia. Technium Soc. Sci. J. (2022). https://doi.org/10.47577/tssj.v36i1.7442
5. Syailendra, M.R., Lie, G., Sudiro, A.: Personal data protection law in Indonesia: challenges and opportunities. Indonesia Law Rev. **14**(2), 175 (2024)

6. Herwanto, G.B., Ekaputra, F.J., Quirchmayr, G., Tjoa, A.M.: Towards a holistic privacy requirements engineering process: Insights from a systematic literature review. IEEE Access (2024)
7. Herwanto, G.B., Nurwidyantoro, A., Ningtyas, A.M., Nurfajri, M.O., Quirchmayr, G., Tjoa, A.M.: Privacy patterns and objectives for legally compliant software based on the Indonesia's PDP law. In: Pardede, E., Ma, Q., Kotsis, G., Amagasa, T., Nadamoto, A., Khalil, I. (eds.) Information Integration and Web Intelligence. iiWAS 2025. LNCS, vol. 16330, pp. 266–281. Springer, Cham (2026). https://doi.org/10.1007/978-3-032-11976-6_19
8. Putri, N.M.D.G., Girinatha, D.G.W.: Legal protection of personal data of Indonesian citizens based on act number 27 of 2022. Jurnal Notariil (2024). https://doi.org/10.22225/jn.9.2.2024.71-75
9. Utomo, S.D.: Personal data protection through law number 27 of 2022: challenges of cybercrime in the era of globalization and digital. Pena Justisia: Media Komunikasi Dan Kajian Hukum (2024). https://doi.org/10.31941/pj.v23i3.5611
10. Al-Momani, A., et al.: Land of the lost: privacy patterns' forgotten properties: enhancing selection-support for privacy patterns. In: Proceedings of the 36th Annual ACM Symposium on Applied Computing, pp. 1217–1225 (2021)
11. Bumpenboon, T.: Thailand's personal data protection act: an understanding from the perspectives of the European privacy law. Thammasat Rev. Econ. Soc. Policy **6**(1), 50–82 (2020). https://doi.org/10.14456/tresp.v6i1.249265

Comparative Evaluation of Machine Learning and Deep Learning Models for Cryptocurrency Price Prediction Under Different Normalization Schemes

Jantima Polpinij(✉), Vuttichai Vichianchai, Theeraya Utha, Anirut Chottanom, Jatuphum Juanchaiyaphum, Thananchai Khamket, and Chumsak Sibunruang

Faculty of Informatics, Mahasarakham University, Kantharawichai 44150, Mahasarakham, Thailand
{jantima.p,vuttichai.v,theeraya.u,anirut,jatuphum.j, thananchai.k,chumsak.s}@msu.ac.th

Abstract. Precise cryptocurrency price prediction is still a difficult problem due to huge volatility, non-stationarity, and scale differences among different assets. Although many works use machine learning (ML) and deep learning (DL) models, the impact of data normalization has not drawn much attention and is regarded as a preprocessing step instead of an experimental factor. This work sets a thorough comparative standard, systematically exploring the interaction between normalization methods and predictive models for short-term cryptocurrency price prediction. Daily closing prices of Bitcoin, Ethereum, and Ripple are normalized using various normalization techniques such as Min–Max scaling, Z-score normalization, and Robust Scaling. A summary of machine learning and deep learning models is compared by forecasting horizon using mean absolute error (MAE), mean absolute percentage error (MAPE), and root mean square error (RMSE) metrics along with computational aspects. The experimental results confirm that normalization plays an important role in the performance and stability of models, regardless of volatility levels across assets. Deep learning architectures, and particularly the GRU gated models, achieve superior prediction accuracy, whereas tree-based methods present competitive results with lower computational demands. The results can serve as a practical reference for model selection in cryptocurrency markets.

Keywords: Cryptocurrency Price Prediction · Data Normalization · Machine Learning · Deep Learning · Time-Series Forecasting

1 Introduction

Cryptocurrency markets have attracted the interest of investors, financial firms, and academics owing to their decentralized nature, high levels of volatility, as well as non-stop trading activity [1, 2]. That is why precise cryptocurrency price prediction is so critical for successful investment decision-making, risk control, and market research. But the prices of cryptocurrencies, unlike those of fiat assets, are significantly nonlinear

S. Sodsee et al. (Eds.): IC2IT 2026, LNNS 5590, pp. 111–123, 2026.
https://doi.org/10.1007/978-3-032-28048-0_11

and highly sensitive to market sentiment, causing volatile price movements and thereby making forecasting difficult [3–5].

Over the last couple of years, using Machine Learning (ML) and Deep Learning (DL) for cryptocurrency price prediction has become very popular. Conventional ML algorithms like Support Vector Regression (SVR) and Random Forest (RF) have shown excellent ability in nonlinear relationship modeling, with relatively low computational cost [6]. With limited data size or the need for fast training, these models are generally preferred [7]. On the other hand, deep learning models, particularly RNNs such as Long Short-Term Memory (LSTM) networks and Gated Recurrent Units (GRU), have demonstrated promising results in handling temporal dependencies inherent in financial time-series data [8–10]. Due to their capacity to capture iterative patterns, they are appealing for predicting cryptocurrency prices in short and medium time frames [11].

Although an increasing number of studies have explored machine learning (ML) and deep learning (DL) approaches for cryptocurrency price prediction, data preprocessing—particularly normalization—has received relatively limited attention [6]. Cryptocurrency price time series represent high-frequency data characterized by large-scale variations and extreme values [12]. Normalization is therefore essential for stabilizing model training, improving numerical robustness, and enabling fair comparisons across different learning algorithms [13, 14]. However, most existing studies employ only a single normalization technique, typically Min–Max scaling or Z-score standardization, without systematically investigating how the choice of normalization method affects predictive performance [6, 15, 16]. Moreover, normalization is generally treated merely as a preprocessing step rather than as an experimental factor; consequently, the interaction between normalization methods and different model types remains largely unexplored [6].

Another drawback in the current literature is the concentration on prediction performance only [17, 18]. Although metrics like MAE and RMSE are widely reported, computational efficiency (both during training and inference) is seldom considered [17]. This limitation diminishes the applicability of many of the proposed models, especially in realistic settings when computational resources are scarce or prediction needs to occur close to real time [19, 20]. Hence, there is still a lack of empirical evidence to help guide practitioners in the search for an optimal trade-off between prediction accuracy and computational cost when choosing ML or DL models [21, 22].

Motivated by the above observations, the objective of this work is to quantitatively analyze how different normalization schemes influence the performance of cryptocurrency price prediction functions and whether these influences are distinct between machine learning models and deep learning models [15, 16]. The motivation of the study is twofold: first, whether some normalization methods are better suited for particular families of models, and second, how much the accuracy improvements achieved by deep learning models justify their increased computational cost. Through the combination of predictive power and computational efficiency, this paper discusses an essential yet less studied issue in cryptocurrency forecasting.

The aim of the present study is, thus, to establish a solid comparative benchmark for future research that particularly emphasizes normalization as an important experimental dimension. By applying the models to the corresponding daily prices of Bitcoin,

Ethereum, and Ripple, it compares benchmark ML and DL approaches across different normalization strategies and evaluates them using mean absolute error (MAE), mean absolute percentage error (MAPE), as well as root mean-square error (RMSE), together with recorded training and prediction times. The main contributions of this work are empirical comparisons demonstrating the interplay between normalization approaches and prediction models, the trade-off between accuracy and efficiency, as well as useful recommendations for choosing accurate yet deployable predictive models in the cryptocurrency market.

The structure of this paper is organized as follows. Section 2 provides a description of the dataset used in this study. Section 3 outlines the research methodology. Section 4 presents and discusses the experimental results. Finally, Sect. 5 summarizes the conclusions.

2 Dataset Description

In this study, historical prices of cryptocurrencies have been collected from the Yahoo Finance website, which is one of the most common sources of data used in financial forecasting research work as it provides wide access to data repositories. The data include daily closing prices of three popular cryptocurrencies: Bitcoin (BTC), Ethereum (ETH), and Ripple (XRP). These cryptocurrencies were included as they capture a variety of market capitalizations, levels of volatility, and adoption quantities, thus providing a diverse test set for predictive models.

The data were retrieved on 1 June 2025, with a continuous historical time span ranging from 2015 to 2023. For each of these cryptocurrencies, the dataset contains standard market attributes available from Yahoo Finance: Open, High, Low, Close, and Adjusted Close prices along with Trading Volume. The closing price is selected as the predicted target since it is prevalent in cryptocurrency price prediction and defines the last consensus of the market during each trading day.

All cryptocurrencies are standardized to the same temporal resolution (daily sampling), enabling consistent and fair comparisons across different cryptocurrencies and predictive models. Prior to model training, the data are merged based on the date attribute, and records with missing values are removed to prevent anomaly-induced bias. The resulting datasets are then normalized using multiple normalization methods, which are systematically evaluated as experimental factors rather than being treated merely as preprocessing techniques (Table 1).

Table 1. Summary of Cryptocurrency Datasets Used in This Study.

Cryptocurrency	Symbol	Number of Data Points
Bitcoin	BTC	3,287
Ethereum	ETC	3,077
Ripple	XRP	3,287

3 Research Method

One of the main drawbacks in the cryptocurrency forecasting literature is that model comparisons are usually performed under non-identical experimental pipelines. Differences in forecasting horizons, data splitting approaches applied, preprocessing, and hyperparameter tuning prevent fair comparison, and reported performance gains could be due to different methodologies rather than superior model capability. Especially, the importance of data normalization is often undervalued but is essential for training stability and prediction accuracy in highly volatile cryptocurrency markets.

To solve this problem, the current work applies a single and leakage-free experimental pipeline that is fully standardized across the entire time series of daily prices of Bitcoin (BTC), Ethereum (ETH), and Ripple (XRP). All models are analyzed in a common forecasting context with the same data split and tuning process, so that differences observed between models can be compared in terms of real-world behavior rather than experimental noise. In contrast to the majority of existing work, normalization is posited as a direct experimental dimension, and its interaction with various learning regimes is investigated.

In this context, the contribution also aims to provide an empirical benchmark of how these normalization strategies influence machine learning and deep learning models in terms of not only predictive accuracy but also computational efficiency. Performance is measured in terms of MAE, MAPE, RMSE, training time, and inference time—providing insights into the accuracy–efficiency trade-off and practical recommendations for choosing accurate yet deployable cryptocurrency forecasting models.

It provides a step-by-step methodology for developing predictive models to predict the closing prices of cryptocurrencies such as BTC, ETH, and XRP. Let $\{p_t\}_{t=1}^{T}$ represent the time series of daily closing price for a cryptocurrency (BTC, ETH, or XRP), and $p_t > 0$. Predictive task is multi-horizon regression that follows the same workflow across all datasets.

3.1 Study Plan: Multi-Horizon Forecasting

Three prediction horizons are studied to represent various decision-making scenarios. 7-day horizon ($H = 7$) instead fosters short-term tactical decisions. As different horizons change the signal-to-noise properties of cryptocurrency prices, we examine whether the relationship between normalization procedures and model types are similar or change with forecast horizon.

The series of cryptocurrency prices is converted into a supervised learning problem by having lookback window length L. For each time step t, the input vector and target prediction are described by:

$$X_t = \left[p_{t-L+1}, p_{t-L+2}, \ldots, p_t\right] \in \mathbb{R}^L, \quad y_t^{(H)} = p_t + H \tag{1}$$

Let p_t present the closing price of a given cryptocurrency at time step t. The vector X_t denotes the input feature vector at time t, constructed using a lookback window of length L, where L is the number of past observations included as model inputs. The target variable $y_t^{(H)}$ corresponds to the closing price at a future time step $t + H$, with H denoting the forecasting horizon. The notation $\mathbb{R}^L$ indicates that the input vector lies in an L-dimensional real-valued space.

It is note that Open, High, Low, Close, and Trading Volume (OHLCV) data are available, additional engineered features (e.g., returns, volatility, or volume changes) may be included, provided that the identical feature set is used across all assets and models.

3.2 Reliable Train/Test Splitting

Given a cryptocurrency and forecasting horizon H, for that given combo of parameters, the data set is divided chronologically into three parts: 70% is devoted to training, 15% validation and 15% testing. There is no shuffling of data, as we take care to evaluate each model in a realistic way: training models on previous observations and evaluating based on next observations. A walk-forward (rolling-origin) validation is used to prevent a "*lucky split*". Folds $k = 1,..., K$ are realized such that in each fold k training is carried out on $[1 : t_k]$ and validation on $[t_k + 1 : t_k + \mathrm{m}]$ after which the origin is forward-rolled. Results are expressed as mean $\pm$ standard deviation (STD) over the folds. All transformations, including normalization, are fitted on the training data of each fold to prevent information leakage.

3.3 Data Normalization

Normalization is considered as a controlled factor. For each fold and normalization method *Nj*, compute transform on training only, apply to validation/test. Let x be a raw feature (e.g., price or return). Let T be training indices.

Min–Max Scaling [16] – This normalization technique transforms the value p_t of the cryptocurrency feature values into a fixed range, similar to [0, 1], based on the minimum and maximum values found in the training data. It is defined as

$$\tilde{p}_t = \frac{p_t - \min(p)}{\max(p) - \min(p)\prime} \tag{2}$$

where let *min(p)* and *max(p)* be the minimum and maximum cryptocurrency's feature values in the training set. This method preserves the relative magnitude of features' value but is sensitive to extreme values, which are common in cryptocurrency markets.

Z-score Standardization [16] - This is a method that turns the cryptocurrency's feature values into having zero mean and unit variance. This would let models to learn on a standard scale. It is defined as:

$$\tilde{p}_t = \frac{p_t - \mu}{\sigma} \tag{3}$$

where μ and σ denotes the mean and standard deviation of the cryptocurrency's feature values computed from the training data. This normalization is widely used in both machine learning and deep learning models to stabilize training and improve numerical convergence.

Robust Scaling [23] – This normalization technique uses robust statistics to normalize the cryptocurrency's feature values, which makes it less affected by outliers and unexpected price surges. It is defined as

$$\tilde{p}_t = \frac{p_t - median(p)}{IQR(p)} \tag{4}$$

where *median(p)* denotes the median of cryptocurrency's feature values and $IQR(p) = Q3 - Q1$ is the interquartile range computed from the training data. This approach is particularly suitable for highly volatile cryptocurrency time series.

3.4 Predictive Model Development

To investigate the interplay between different normalization schemes and families of predictive models, five predictive models from two classes are posterized: classical machine learning (ML) and deep learning (DL). All of them are learned by the identical supervised learning form and tested under same scenarios.

Support Vector Regression (SVR) with RBF Kernel [24] - The SVR served as a nonlinear machine learning baseline because of its great generalization ability and acceptable time cost. The SVR learns a regression function by mapping input samples to a high dimension space, and the complexity of the model is regulated through margin maximization. The regression model is expressed as:

$$f_x(x) = \sum_{i=1}^{n} \left(\alpha_i - \alpha_i^*\right) K(x_i + x) + b \tag{5}$$

where $K(x_i + x) = \exp\left(-\gamma x_i - x^2\right)$ represents the RBF kernel, α_i and α_i^* are Lagrange multipliers and b is a bias term. In our study, we set hyperparameters as $C = 10$, $\gamma = 0.1$, $\epsilon = 0.01$ to trade off prediction accuracy and computations burden. SVR is highly fitting to the forecasting tasks which serve for deployment, and it is also a proper baseline algorithm in revealing the relationship among normalization strategies and the performance of prediction model.

Random Forest (RF) [25] – RF is an ensemble learning algorithm that combines multiple regression trees generated on bootstrapped samples and makes predictions by averaging output of each tree, which in turn reduces variance and enhances generalization. The prediction of the ensemble is expressed as:

$$\hat{y} = \frac{1}{B} \sum_{b=1}^{B} T_b(x) \tag{6}$$

where T_b is the b-th regression tree and B is the size of the forest. In this work, the hyperparameters are set by 200 trees and maximum tree depth of 10 minimum samples per leaf equal to 5 for striking a balance between model expressiveness and control on overfitting. It has been reported that RF is noise-robust and robust to scale effects, resulting in a powerful machine learning baseline that generalize well across different normalization schemes, and thus it is suitable for comparative benchmarking.

Gradient Boosting Regressor (*XGBoost*) [26] - In XGBoost, framed decision trees are added sequentially and each new tree is fit to the residual errors of the current model. The iteration m prediction is given by:

$$\hat{y}^{(m)} = \hat{y}^{(m-1)} + \eta f_m(x) \tag{7}$$

where $f_m(x)$ represents the newly fitted regression tree and η is a learning rate that manages the contribution of each tree. The model is set with 300 trees, a learning rate of 0.05, the depth of each tree is bounded by 6 in this study. The XGBoost model are also a strong contemporary tabular-learning baseline- high predictive power, but computationally lightweight and lend themselves well to comparison in tabular benchmarks which do no0t use the same normalization.

Long Short-Term Memory (*LSTM*) [27] - LSTM is a type of recurrent neural network architecture that has been developed so as to capture long-range temporal dependencies in sequential data using gated memory units, where updating the cell states and computing the hidden states are given by:

$$c_t = f_t \odot c_{t-1} + i_t \odot \tilde{c}_t, h_t = o_t \odot \tanh(c_t) \tag{8}$$

where i_t, f_t and o_t are the input, forget and output gates respectively. For our LSTM model in this work, we used 64 hidden units and a dropout rate of 0.2 and train the network for 30 epochs with batch size of 32, using Adam optimizer. LSTM in fact is considered a baseline of deep learning algorithms for forecasting cryptocurrency price, and designed to capture the temporal dependence towards different prediction horizons.

Gated Recurrent Unit (*GRU*) [28] - GRU is a simplified recurrent neural networks architecture which integrate memory gating mechanisms into update and reset gates resulting in reduced model complexity while retaining the ability for temporal modeling. The update of hidden state is controlled by:

$$h_t = (1 - z_t) \odot h_t + z_t \odot \tilde{h}_t \tag{9}$$

where z_t is the update gate. The GRU model has 64 hidden units, a dropout rate of 0.2 and is trained using a batch size of 32 and for up to 30 epochs. We use GRU to show the tradeoff between predictive accuracy and computational efficiency, because it generally gives the same performance compare with LSTM while reduces training and inference complexity.

4 Evaluation

4.1 Evaluation Metrics

Model performance is measured using the Mean Absolute Error (MAE), the Mean Absolute Percentage Error (MAPE) and Root Mean Square Error (RMSE) in combination with training and prediction timing to judge computational efficiency. MAE measures average absolute deviations,

$$MAE = \frac{1}{N}\sum_{i=1}^{N} |y_i - \hat{y}_i| \tag{10}$$

while MAPE provides scale-independent relative error comparison across cryptocurrencies,

$$MAPE = \frac{100}{N} \sum_{i=1}^{N} \left| \frac{y_i - \hat{y}_i}{y_i} \right| \quad (11)$$

RMSE penalizes large forecasting errors and is defined as

$$RMSE = \sqrt{\frac{1}{N} \sum_{i=1}^{N} (y_i - \hat{y}_i)^2} \quad (12)$$

Furthermore, the training and inference times are studied to measure model efficiency. These metrics in combination provide a balanced assessment of predictive accuracy and computational cost across various normalization schemes and model architectures.

4.2 Results and Discussion

In this section, we provide a detailed comparison of machine learning and deep learning models for cryptocurrency price prediction on BTC, ETH, and XRP using different normalization schemes. MAE, MAPE, and RMSE are computed for all models along with training time and inference time to assess predictive accuracy as well as efficiency. The findings provide empirical evidence of how normalization schemes interact with different model families and how such effects change across cryptocurrencies exhibiting different volatility behavior. The results are presented in Table 2.

Empirically, we find that the errors for XRP are almost always significantly lower in absolute terms (MAE, RMSE) than those of BTC, with ETH being somewhere in between. This is due to the scale and price level of the assets involved. Notice that Bitcoin and Ether are exchanged at much higher levels compared to XRP, resulting in larger absolute errors even if the relative prediction accuracy is comparable. Compared to them, XRP's price is at a lower scale of value, and therefore the absolute deviations are also smaller in terms of MAE and RMSE.

Also, XRP possesses 7-day short-term price dynamics that are smoother than those of BTC and less significant in medium-term movements. Bitcoin is subject to sudden increases driven by changes in macroeconomic fundamentals, regulatory news, and large transactions, the latter exacerbating forecast errors and thus short-run volatility. Although the relative error (MAPE) for XRP is higher than BTC in some configurations, its absolute error values are consistently low; therefore, it can be concluded that even though we may have graph topology which constrains the bounds of prediction power, overall both time series exhibit weaker statistical relationships.

Across all three cryptocurrency types, we observe that Z-score normalization and Robust Scaling outperform Min–Max scaling. Min–Max normalization is influenced by outliers, and extreme values are frequent in cryptocurrency markets because of sudden price bubbles. Such sensitivity results in constrained feature distribution in the presence of outliers, which undermines the efficacy of learning interpretable temporal patterns.

Z-score normalization shows more stable performance, as it focuses on centering the data and standardizing variance for both linear and non-linear models. These effects are more pronounced for BTC and ETH, where volatility is higher and the distribution of prices has heavy tails. The Robust Scaling variant boosts performance by using a structure based on the median and interquartile range, which is well adapted to datasets with frequent anomalies or extreme fluctuations. The performance difference between the normalization methods is closer for XRP, which is consistent with its lower volatility compared to other cryptocurrencies in a 7-day horizon.

Table 2. Experimental Results for the 7-Day Predicting Horizon.

Crypto Type	Normalization	Models	MAE	MAPE	RMSE
BTC	Min-Max	SVR	820	2.15	1,040
		RF	760	1.98	980
		XGBoost	710	1.82	930
		LSTM	660	1.65	880
		GRU	640	1.60	860
	Z-score	SVR	780	20.5	1,000
		RF	720	1.88	940
		XGBoost	670	1.72	900
		LSTM	620	1.55	850
		GRU	600	1.50	830
	Robust Scaling	SVR	760	2.00	980
		RF	700	1.85	920
		XGBoost	650	1.68	880
		LSTM	600	1.52	830
		GRU	580	1.48	810
ETH	Min-Max	SVR	62	2.60	78
		RF	58	2.35	74
		XGBoost	54	2.10	70
		LSTM	50	1.95	66
		GRU	48	1.90	64
	Z-score	SVR	59	2.45	75
		RF	55	2.20	71
		XGBoost	51	2.00	67
		LSTM	47	1.80	61
		GRU	45	1.80	61
	Robust Scaling	SVR	57	2.40	73

(*continued*)

Table 2. *(continued)*

Crypto Type	Normalization	Models	MAE	MAPE	RMSE
		RF	53	2.15	69
		XGBoost	49	1.95	65
		LSTM	45	1.82	61
		GRU	43	1.78	89
XRP	Min-Max	SVR	0.038	3.1	0.048
		RF	0.035	2.85	0.045
		XGBoost	0.032	2.55	0.042
		LSTM	0.029	2.35	0.039
		GRU	0.028	2.30	0.038
	Z-score	SVR	0.036	2.95	0.046
		RF	0.033	2.70	0.043
		XGBoost	0.030	2.45	0.040
		LSTM	0.027	2.25	0.037
		GRU	0.026	2.20	0.036
	Robust Scaling	SVR	0.035	2.90	0.042
		RF	0.032	2.65	0.042
		XGBoost	0.029	2.40	0.039
		LSTM	0.026	2.22	0.036
		GRU	0.025	2.18	0.035

Notably, in terms of classical machine learning models, SVR always obtains the worst performance on all datasets and normalization approaches. SVR does not handle complex nonlinear temporal relationships well, and performance may be subject to kernel and hyperparameter selection, particularly in highly volatile financial time series.

Random Forest (RF) is an improvement over SVR since it can model complex, non-linear relationships through ensemble learning; however, its effectiveness for time-series prediction tasks is restricted as it does not model time explicitly. XGBoost systematically surpasses RF and SVR, which is due to gradient boosting as well as sequential error compensation, especially when paired with Z-score or Robust Scaling.

LSTM and GRU are the best deep neural networks for all cryptocurrencies. Their better accuracy comes from their capacity to capture both long- and short-term temporal connections, especially for Bitcoin and Ethereum. GRU usually performs slightly better than LSTM, probably because its gating mechanism is simpler, which helps avoid over-fitting and improves generalization when there is limited training data. The performance improvements of LSTM and GRU are especially noticeable when used with Robust Scaling. This shows that normalization methods and recurrent architectures work well together.

In general, the experimental results show that the properties of the dataset, the normalization procedures, and the model architectures all have an effect on predictive performance. Price scale has a big effect on absolute error measurements, which is why the error levels for Bitcoin are much higher than those for Ripple. Robust Scaling and Z-score normalization both make models more stable and accurate consistently, although they work best on time series or data with outliers. Deep learning models, notably GRU, are clearly better than traditional machine learning methods when it comes to predicting cryptocurrency prices in the short term.

These results show how important it is to regard normalization as an experimental factor instead of just a preprocessing step. They also show that predictive models cannot be accurately judged without considering the data's characteristics and normalization choices. Deep learning models are frequently very accurate, but tree-based methods like XGBoost work almost as well at a much lower cost, making them suitable for situations where both training and inference speed are important.

These results create a common standard for comparing cryptocurrencies, normalization procedures, and model families. This will help future research and real-world applications.

5 Conclusion

This paper provided a systematic comparison of ML and DL models for predicting cryptocurrency prices in the short term, where the significance of data normalization was emphasized as an important experimental dimension. For the datasets from Bitcoin, Ethereum, and Ripple (at the daily level), a few normalization techniques were considered along with several classifiers at different forecasting horizons. The findings indicate that normalization has a significant effect on predictive power and model stability and should not be treated as a mere preprocessing routine.

Our empirical findings show that Robust Scaling and Z-score normalization outperform Min–Max scaling significantly, particularly for highly volatile assets. Deep learning models, in particular gated recurrent units (GRUs), offer better accuracy by capturing time dependencies well, but tree-based models like XGBoost strike a good balance between accuracy and computational efficiency. In addition, the absolute error-based measures are heavily dependent on the price scale of the asset, demonstrating the need to compare results based on their data properties.

In conclusion, this work sets up a strong comparative benchmark that demonstrates the interaction of normalization techniques with model architecture and provides useful information on how to select reliable yet deployable predictive models for the cryptocurrency market. This framework could be further applied to longer-term horizons, additional assets, and real-time settings.

Acknowledgement. This work was financially supported by Mahasarakham University.

References

1. Iqbal, R., Riaz, M., Sorwar, G., Qadir, J.: Cryptocurrency market volatility and forecasting: A comparative analysis of modern machine learning models for cryptocurrencies predicting accuracy. Rev. Pac. Basin Financ. Markets Policies **27**(4) (2025)
2. Alqaralleh, H., Abuhommous, A.A., Al-Saraireh, A.: Modelling and forecasting the volatility of cryptocurrencies: a comparison of nonlinear GARCH-type models. Int. J. Financ. Res. **11**(4) (2020)
3. Hossain, M.F.B., Lamia, L.Z., Rahman, M.M., Khan, M.M.: FinBERT-BiLSTM: a deep learning model for predicting volatile cryptocurrency market prices using market sentiment dynamics (2024). arXiv preprint arXiv:2411.12748
4. Zubair, M., Ali, J., Alhussein, M.A., Hassan, S., Aurangzeb, K., Umair, M.: An improved machine learning-driven framework for cryptocurrencies price prediction with sentimental cautioning. IEEE Access **12** (2024)
5. Pang, Y., Sundararaj, G., Ren, J.: Cryptocurrency price prediction using time series and social sentiment data. In: Proceedings of the 6th IEEE/ACM International Conference on Big Data Computing, Applications and Technologies, pp. 35–41. IEEE, New York (2019)
6. Murray, K., Rossi, A., Carraro, D., Visentin, A.: On forecasting cryptocurrency prices: a comparison of machine learning, deep learning, and ensembles. Forecasting **5**(1) (2023)
7. Hegde, B.P., Roopa, U.: A comprehensive study on comparison of long short-term memory, support vector machine, and their hybrid model performance using erratic cryptocurrency data. In: ITM Web of Conferences (2024)
8. Nair, M., Marie, M.I., Abd-Elmegid, L.A.: Prediction of cryptocurrency price using time series data and deep learning algorithms. Int. J. Adv. Comput. Sci. Appl. **14**(8) (2023)
9. Patel, M., Tanwar, S., Gupta, R., Kumar, N.: A deep learning-based cryptocurrency price prediction scheme for financial institutions. J. Inf. Secur. Appl. (2024)
10. Dutta, A., Kumar, S., Basu, M.: A gated recurrent unit approach to bitcoin price prediction. J. Risk Financ. Manag. **13**(2) (2019)
11. Kim, J., Kim, S., Wimmer, H., Liu, H.: A cryptocurrency prediction model using LSTM and GRU algorithms. In: Proceedings of the International Conference on Big Data, Cloud Computing, Data Science & Engineering (2021)
12. Simtharakao, S., Sutivong, D.: Exploring normalization techniques in neural networks for bitcoin candlestick price prediction. In: Digital Signal Processing and Signal Processing Education Workshop (2023)
13. Lima, F.T., Souza, V.M.A.: A large comparison of normalization methods on time series. Big Data Res. **34** (2023)
14. Cabello-Solorzano, K., Ortigosa de Araujo, I., Peña, M., Correia, L., Tallón-Ballesteros, A.J.: The impact of data normalization on the accuracy of machine learning algorithms: a comparative analysis. In: García Bringas, P., et al. (eds.) 18th International Conference on Soft Computing Models in Industrial and Environmental Applications (SOCO 2023). SOCO 2023. LNNS, vol. 750, pp. 344–353. Springer, Cham (2023). https://doi.org/10.1007/978-3-031-42536-3_33
15. Polpinij, J., Saktong, N.: A comparative study of machine learning approaches for predicting close-price cryptocurrency. In: Proceedings of the International Conference on ICT and Knowledge Engineering, Bangkok (2022)
16. Phasook, C., Polpinij, J., Luaphol, B.: A study of comparative methods for closed-price cryptocurrency prediction. In: Proceedings of the International Conference on Information Technology, Bangkok (2022)
17. Dehghani, M., Arnab, A., Beyer, L., Vaswani, A., Tay, Y.: The efficiency misnomer. In: Proceedings of the International Conference on Learning Representations (ICLR) (2021)

18. Al-Jarrah, O.Y., Yoo, P., Muhaidat, S., Karagiannidis, G., Taha, K.: Efficient machine learning for big data: a review. Big Data Res. **2**(3), 87–93 (2015)
19. Sun, P., Aljeri, N., Boukerche, A.: Machine learning-based models for real-time traffic flow prediction in vehicular networks. IEEE Netw. **34**(3), 178–185 (2020)
20. Roth, W., et al.: Resource-efficient neural networks for embedded systems (2020). arXiv preprint arXiv:2006.05656
21. Törnquist, E., Santos, W.C., Pogue, T.C., Wingle, N., Caulk, R.: Balancing computational efficiency and forecast error in machine learning-based time-series forecasting: insights from live experiments on meteorological nowcasting (2023). arXiv preprint arXiv:2309.15207
22. Yang, B., Gül, M., Chen, Y.: Comparative analysis of deep learning and tree-based models in power demand prediction: accuracy, interpretability, and computational efficiency. J. Build. Phys. **49**(1) (2025)
23. Bae, G., Kim, J.: Observing cryptocurrencies through robust anomaly scores. Entropy **24**(11) (2022)
24. Basak, D., Pal, S., Patranabis, D.C.: Support vector regression. Neural Inf. Process. – Lett. Rev. **11**(10), 203–224 (2008)
25. Svetnik, V., Liaw, A., Tong, C., Culberson, J., Sheridan, R., Feuston, B.: Random forest: a classification and regression tool for compound classification and QSAR modeling. J. Chem. Inf. Comput. Sci. **43**(6) (2003)
26. Fafalios, S., Charonyktakis, P., Tsamardinos, I.: Gradient boosting trees. In: Encyclopedia of Machine Learning and Data Mining. Springer, Berlin, Heidelberg (2020)
27. Sak, H., Senior, A., Beaufays, F.: Long short-term memory recurrent neural network architectures for large scale acoustic modeling. In: Proceedings of Interspeech. ISCA (2014)
28. Chung, J., Gülçehre, Ç., Cho, K., Bengio, Y.: Empirical evaluation of gated recurrent neural networks on sequence modeling (2014). arXiv preprint arXiv:1412.3555

Impact of Working Hours and Overtime Decisions on Makespan in Job Shop Scheduling

Jaramporn Hassamontr(✉), Wannaporn Suwannarat, and Bunthita Rasniyom

King Mongkut's University of Technology North Bangkok, Bangkok 10800, Thailand
jaramporn@gmail.com, swannaporn58@gmail.com,
banthitarasniyom@gmail.com

Abstract. The purpose of this research is to develop a simulation tool to investigate how working hour and overtime decisions affect overall schedule performances. Specifically, we are interested in traditional performance indicator, makespan in job shop scheduling environments, as well as cost factor represented by overtime hours. The tool developed can be used to provide tradeoffs for companies looking to speed up the work while keeping overtime cost under control. A benchmark test case available in literature is used as a case study to illustrate the tool's capabilities.

Keywords: Job Shop Scheduling · Makespan · Working Hours · Overtime Decision

1 Introduction

Speed is a critical competitive advantage in most businesses. This is especially true for companies manufacturing highly customized products such as machines and tooling. As customers demand different product configurations to meet their needs, it is difficult for such companies to standardize their production lines. Instead, most are organized as a job shop where each job has a unique process routing, going through various machines or process steps in specific sequence. Since the number of machines in a job shop is limited, and there are many ongoing jobs at the same time, this causes job's routings to get entangled, leading to long queue in some machines and idle time in others.

This type of problem is known as job shop scheduling which has been studied by researchers since 1950s (Jain [1]). Researchers attempted to determine job sequence in each machine along with resulting operation start and finish times so that the last job finishes its last operation at the soonest possible time. This finish time of the last operation of the last job is a well-known performance indicator, called makespan. This problem is NP hard (Jain [1]).

In solving job shop scheduling problems, most researches focused on sequencing operations into machines without considering constraints from resources. To illustrate the rationale behind our research, let us consider an example of a job schedule presented by Gantt chart in Fig. 1(a). This chart represents the best-known solution (Shylo [2])

S. Sodsee et al. (Eds.): IC2IT 2026, LNNS 5590, pp. 124–133, 2026.
https://doi.org/10.1007/978-3-032-28048-0_12

from a benchmark problem widely available in the literature, TA73 (Taillard [3]). Time is shown on the x-axis in minutes. Job sequence in each machine is shown in each row. The first operation in machine 0, for example, is job 21 with its process number 0. This is designated as (21,0) in the figure. This job will then go to process number 1 at machine 3 and so on. Each job has a unique color shading for visualization purposes.

There are two underlying assumptions in this job schedule-(1) work can be interrupted and continued with no effect on product and (2) labor and machines are always available to perform the operation. These two assumptions are not necessarily true in practice. In many manufacturing settings, once machine is setup for a job, it is preferable to run to completion. Once turned off, most machines require additional setup time before resuming work. Also, in terms of resource availability, there are always necessary periodic breaks to maintain machines. This is why most facilities operate in shifts. If these two assumptions are removed, using a work shift of 10 h (600 min), the Gantt chart can be adjusted and shown in Fig. 1(b). So, the real world makespan is indeed larger than the theoretical makespan.

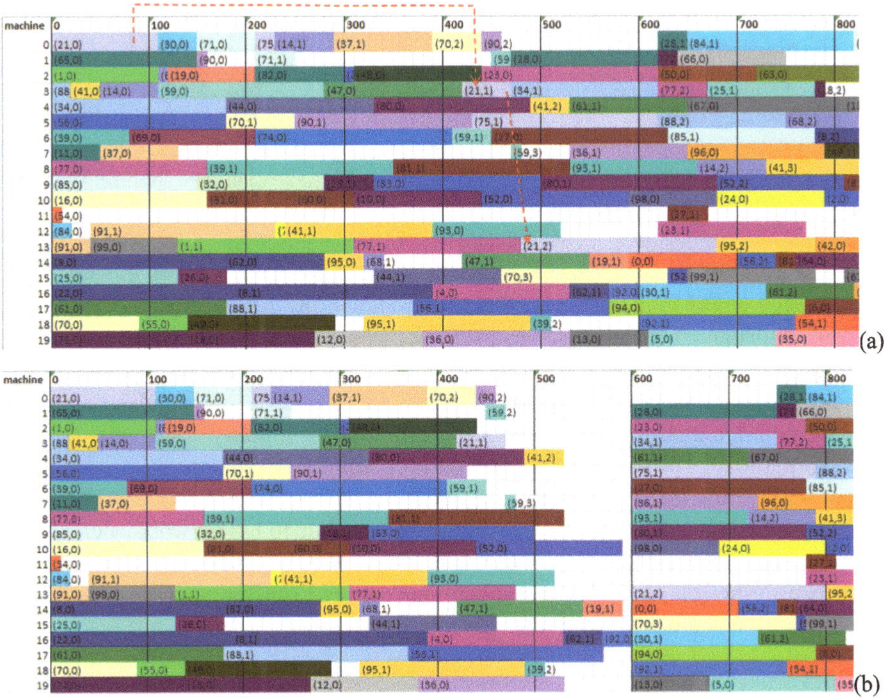

Fig. 1. Comparison of Gantt charts when (a) not considering working hours, and (b) when working in shift of 10 h (600 min).

This research aims to investigate the impact of working hours on makespan. Conceptually increasing working hour in a shift should lead to lower makespan. However, there could be an optimal working hour in terms of cost components. Also, given a working hour format, companies can make overtime (OT) decisions to improve its makespan performance. As a preliminary study, we will develop a simulation tool to re-adjust start-finish times based on a given working hour and overtime decision. Starting job schedule is extracted from best known solutions widely available in the literature. The impact of working hour format and OT combination will be investigated from both time and costs.

2 Literature Review

Job shop scheduling problems were studied by researchers well over 70 years. Jain [1] did quite extensive survey on various approaches to solve the problem with the goal of minimizing makespan. However, as mentioned by Zeng [4], there are three ways to deal with work calendar in job shop scheduling-(1) not considering work calendar at all, (2) considering work calendar only after scheduling and (3) considering work calendar during scheduling. Majority of research work in job shop scheduling fell into the first category since adding work calendar into consideration will make the problem much more difficult to solve. This is especially true when the number of jobs and the number of machines increase. It should be noted, however, several commercial software to perform job shop scheduling, such as Google OR-Tools [5] and IBM CP Optimizer [6], are designed for large-scale problems and do not consider work calendar either. There are some situations where considering work calendar is unnecessary. For one, if one expects resources to operate continuously for a very long time, there is no need to consider work calendar. Job schedule with very large makespan, i.e. spanning through several weeks, is another example.

A limited number of works related to categories 2 and 3 are as follows. Xiong [7] proposed a methodology to map time in the original schedule to work calendar. One of the earlier works to include overtime was Goodwin [8]. Lagodimos [9] investigated tradeoffs between regular shift versus overtime. Artigues [10] proposed a hybrid model to simplify branch-and-bound search to solve job shop scheduling involving employee schedules. Thenarasu [11] attempted to apply the optimization tool in industrial setting, a press shop. Cloutier [12] proposed a methodology to simplify constraints on work calendar in order to be solved by Constraint Programming technique.

Our work falls into the second category as mentioned earlier. We start with a job schedule already available, then we re-adjust the schedule to fit into work calendar. Our main interest is to investigate how working hour and overtime decisions move the overall makespan further out. In order to compare makespans from various cases objectively, one needs to keep the original timeline as a base case.

3 Research Methodology

3.1 Problem Settings

The following assumptions are made in this study.

- Job schedule has already been generated a priori.
- Job sequence in each machine will not be altered. Only its start-finish times are re-calculated based on standard working hours and overtime decisions.

- All machines must follow the same standard working hours and overtime decisions. Overtime on just one machine is not allowed.
- Once an operation is started on a machine, it must continue until completion.
- Standard working hours considered in this study are classified based on a shift of 8-h work. This means the maximum number of shifts per day is three. For one and two shifts, standard working hours are 8 and 16 h respectively. When the company operates 3 shifts per day, they have to stop the work at some point. This leads to a variety of standard working hours. For example, working 5–6 days means standard working hours of 120–144. Standard working hours considered in this study are 8, 16, 120 and 144 h.
- The maximum overtime hours per shift is limited to 4 h. It must be specified in increment of 30 min.

There are 4 types of overtime designation in this study. The first two, *no_OT* and *fixed_OT*, are the most common in industries. Some companies may choose to strictly prohibit overtime, denoted by *no_OT*, while some opt to pre-specify equal overtime hours per shift, say 8 + 2 h per shift, *fixed_OT*. The other two are proposed by the authors, namely minimum overtime, *min_OT* and overtime based on critical machines or critical jobs, *critical_OT*. These two overtime policies will be discussed in more details in Sect. 3.2.

It should be noted that the sum of standard working hours and overtime hours represents the duration that machine can operate continuously. That is, machine needs not to be stopped when there is a shift change as long as it is still in this time duration.

3.2 Simulation Tool Development

In order to study the impact of working hours and overtime on job schedule, a simulation tool was developed on Microsoft Excel using Visual Basic Application. The first step is to decode benchmark problem instance into job routing and solution order into job schedule. Benchmark problem instances were downloaded from Weise [13] whereas the best-known solution can be downloaded from Shylo [2]. These benchmark problems, though not specifying unit for process time, it is safely assumed to be in minutes. And since we are interested in evaluating the impact of working hours above 144 h, that means, the problem instance selected should have makespan well above 144*60 = 8,640 min.

As an example, the benchmark problem called TA73 has best-known makespan of 5,568 min. Each test case series TA7*x* has 20 machines, numbered from 0 to 19, and 100 jobs, numbered from 0 to 99. Each job must visit each machine exactly once according to process number. Again, process number starts from 0 to 99. Since a schedule with makespan larger than 8,640 min is needed, we can simply multiply all process times in all machines for test case TA73 by 2. This will lead to the schedule with makespan of 2*5,568 = 11,136 min. The resulting job routing is shown partially in Fig. 2(a) for job 80. The job starts at machine 4 and finishes at machine 14. The total workload (*TWL*) for all 20 process steps is 2,552 min. After the first process, job 80 will have cumulative workload (*CWL*) of 160 min and has 2,392 min total work remaining (*twkr*). The total work remaining will be used later to identify which job is critical.

The best-known solution is shown as job schedule in Fig. 2(b). It turned out machine 2 is the one in which the last job is completed. That is job 87 finishes at time 11,136 as calculated earlier. Most of the columns in this table are self-explanatory. The arrival time denotes the time of which job arrives the machine. The rightmost column, earliest start time reset, *EST_reset*, will be used when we need to re-calculate start and finish times based on working hours.

job	proc_no	machine	WL (min)	TWL	CWL	twkr
80	0	4	160	2552	160	2392
80	1	9	176	2552	336	2216
80	2	16	114	2552	450	2102
80	3	12	60	2552	510	2042
80	4	19	188	2552	698	1854
80	5	5	190	2552	888	1664
:	:	:	:	:	:	:
80	15	11	130	2552	2088	464
80	16	7	68	2552	2156	396
80	17	8	182	2552	2338	214
80	18	13	188	2552	2526	26
80	19	14	26	2552	2552	0

(a)

machine	job	proc_no	WL (min)	start	finish	arrival	EST reset
2	1	0	112	0	112	0	
2	68	0	12	112	124	0	
2	19	0	86	124	210	0	
2	82	0	94	210	304	0	
2	29	0	8	304	312	0	
2	48	0	130	312	442	0	
2	23	0	178	442	620	0	
2	50	0	98	620	718	0	
2	63	0	134	718	852	0	
:	:	:	:	:	:	:	
2	83	19	126	10868	10994	10636	
2	87	19	136	11000	11136	11000	

(b)

Fig. 2. Supporting tools-(a) job routing, (b) job schedule.

Figure 3 illustrates how schedule is re-adjusted based on specified working hours. When there is no overtime, *no_OT*, as shown in Fig. 3(a), the end of work period (*EOP*) is represented by standard working hours (*std_h*). The algorithm simply looks at each machine to identify if there is a job immediately after or overlapping this *EOP* line. This particular job will be marked to start after time *EOP* only. As an example in Fig. 3(a), the job shaded in green in machine *M*0 and job shaded in blue in machine *M*1 can only start after time *EOP*. This is easily done by writing time *EOP* in the *EST_reset* column of corresponding machine and job in schedule template in Fig. 2(b). It should be noted that in the case of fixed overtime hours, *fixed_OT*, similar approach can be used since one can consider constant overtime hours as part of standard hours.

The third overtime pattern, *min_OT*, is designed to accommodate the job that is being performed during the end of standard hour, *std_h*. The schematics to identify minimum overtime is shown in Fig. 3(b). Here, only the jobs that overlap *std_h* line are considered. In Fig. 3(b), the job with blue shade surpasses the *std_h* line the least. Thus, the overtime will be determined based on the blue-shaded job, rounded to 30 min. The end-of-period will be the sum of standard working hours and rounded overtime hours.

A more rational approach is to determine overtime based on critical job and machine during the end of standard working hours. Here, a critical job is defined as a job with largest total work remaining (*twkr*). A critical machine is also defined as a machine with largest remaining workload (*rwl*). Both definitions look similar but one is for the job and the other is for the machine. To identify both, it is necessary to create a separate note to identify which jobs overlap which machine at *std_h* line.

Figure 4(a) depicts such a note. At time *std_h* of 600 min, machine 0 is not working on any job, thus not shown, while all other machines are occupied. The critical machine is machine 8 where it has 10,446 min of remaining workload, not including the job 93 being performed. Job 93 is expected to finish (*fin_time*) at time 658 min. To help machine 8 finishes the job on hand, one would like to have overtime at least 658–600 = 58 min.

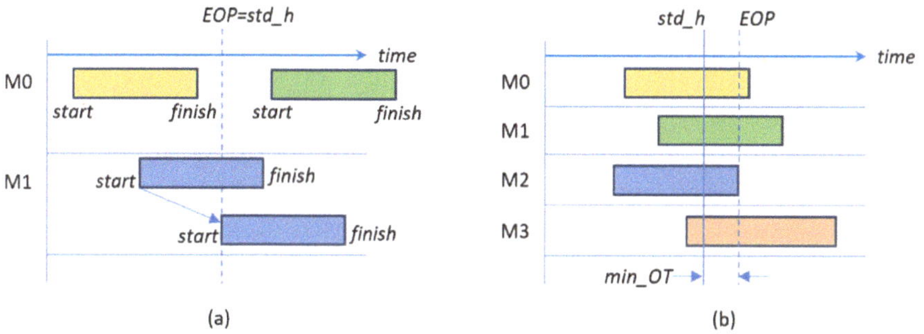

Fig. 3. Identifying the end of shift when (a) no overtime/fixed overtime and (b) minimum overtime is used.

The critical job, among all jobs being performed at time 600, is job 80 on machine 9. Job 80 has 2,216 min of total work remaining, not including the work being done on machine 9. Job 80 is expected to finish at time 678 min, thus an overtime of 678–600 = 78 min should be considered. Our algorithm will use the maximum from the two, which is 78 min and rounded to nearest 30 min, yielding 90 min of overtime. The sum of standard working hour (600 min) and 90 min overtime leads to work period (*EOP*) of 690 min. This *EOP* is then placed on *EST_reset* column in job schedule, shown in Fig. 4(b). Again, only machine 2 partial schedule is shown. All jobs arriving after time *EOP* are to be re-scheduled. Their start, finish and arrival times are deleted and re-calculated.

The re-scheduling process is repeated for each machine and each work period until a new work period cannot be added anymore. The last component of the simulation tool is to generate Gantt chart such as those shown in Fig. 1. The re-scheduling simulation can be validated based on each operation's arrival, start and finish times and also from the Gantt chart generated.

machine	fin_time	rwl	job	proc_no	twkr
1	616	8550	28	0	1998
2	620	9048	23	0	1950
3	616	9444	34	1	1776
4	652	9920	61	1	1636
5	616	10406	75	1	2160
6	630	9676	27	0	1736
7	654	9146	36	1	2020
8	658	10446	93	1	1812
9	678	10046	80	1	2216
10	676	9790	98	0	2044
13	680	9162	21	2	1794
14	702	8392	0	0	1954
15	630	9636	70	3	1496
16	600	9848	92	0	2148
16	728	9720	30	1	1496
17	766	9438	94	0	2126
18	756	9834	92	1	1992
19	610	9386	13	0	1564

(a)

machine	job	proc_no	WL (min)	start	finish	arrival	EST reset
2	1	0	112	0	112	0	
2	68	0	12	112	124	0	
2	19	0	86	124	210	0	
2	82	0	94	210	304	0	
2	29	0	8	304	312	0	
2	48	0	130	312	442	0	
2	23	0	178	442	620	0	
2	50	0	98			0	690
2	63	0	134				
:	:	:	:				
2	83	19	126				
2	87	19	136				

(b)

Fig. 4. (a) Template to help identifying critical job, machine and overtime, (b) clearing schedule beyond end-of-shift in order to reschedule.

4 Results

The simulation tool described in Sect. 3 has been developed and verified. As a preliminary investigation into how standard working hours and overtime decision affect the overall makespan of the system and cost, a case study of test case TA73, with all their process times doubled, is used. The standard working hours considered are 8, 16, 120 and 144 h. There are four overtime patterns studied, namely *no_OT*, *fixed_OT* (1–4 h), *min_OT* and *critical_OT*.

Based on all these combinations, the resulting makespans are shown in left side of Fig. 5(a). As expected, the longer one continues to work, either by increasing the standard working hours or adding overtime, the lower the makespan. Since the minimum overtime one can use in this simulation is 30 min, we can expect *min_OT* policy to perform roughly in between no overtime and 1-h overtime. Critical overtime, on the other hand, does not look to minimize the overtime, thus it tends to perform better than minimum overtime policy. But when *std_h* jumps to 120–144 h, the impact of overtime decision on makespan diminishes.

From makespan result, it is not clear how one policy performs better than the other. In reality, most companies would try to speed up the work with minimum cost. It should be noted that makespan is based on continuous work calendar of the factory. If makespan results are converted into work period, as shown on the right side of Fig. 5(a), the result is much clearer. For standard working hours of 8 and 16, each work period is one day. Working 8 h per day with no overtime, in our case, will require 29.28 days to complete. Using 4-h overtime will reduce the time by 29.28–17.54 = 11.74, almost 12 days. By increasing the standard working hours, the impact of overtime hours is reduced drastically. It is also obvious, *critical_OT* requires less work periods than *min_OT*. And *min_OT* requires less work periods than *no_OT*.

makespan

		fixed_OT				min	crit
std_h	no_OT	1 h	2 h	3 h	4 h	min_OT	crit_OT
8	14056	13696	13558	13380	12630	13940	13610
16	12632	12392	12270	12196	12390	12398	12376
120	11310	11262	11234	11250	11310	11232	11292
144	11244	11278	11306	11300	11276	11274	11274

work periods

	fixed_OT				min	crit
no_OT	1 h	2 h	3 h	4 h	OT	OT
29.28	25.36	22.60	20.27	17.54	28.10	24.10
13.16	12.15	11.36	10.70	10.33	12.63	11.83
1.57	1.55	1.53	1.52	1.52	1.56	1.56
1.30	1.30	1.29	1.28	1.27	1.30	1.30

(a)

OT hours

	fixed_OT				min	crit	max
std_h	1 h	2 h	3 h	4 h	OT	OT	OT
8	25	44	60	68	8.5	35.0	3.5
16	12	22	30	40	4.5	17.0	3.0
120	1	2	3	4	0.5	1.5	1.5
144	1	2	3	4	0.5	0.5	0.5

Total Cost

	fixed_OT				min	crit
no OT	1 h	2 h	3 h	4 h	OT	OT
234.27	240.40	246.77	252.16	242.32	237.58	245.33
210.53	212.38	214.78	216.17	225.20	208.88	214.77
188.50	187.65	187.16	187.43	188.42	187.45	188.95
187.40	188.17	188.85	188.99	188.85	188.15	188.15

(b)

Fig. 5. Resulting makespan and number of shifts.

The left side of Fig. 5(b) demonstrates total overtime hours in each case. Fixed overtime, as the name suggests, will depend on how many periods are used. Minimum overtime and critical overtime shown in the Figure are cumulative values collected from all work periods. Furthermore, since critical overtime is intended to clear critical machine and job without considering overtime limit, its maximum overtime, at any one work period, is also reported. In the case of *std_h* = 8 h, *critical_OT* requires a total overtime hour of 35 over 24 work periods, averaging 1.458 h per work period, even though the maximum overtime from one period reaches 3.5 h.

Work period and overtime pay can be used to identify trade-off. To properly represent overtime cost, let us define an overtime cost factor based on labor cost. In Thailand, for example, labor wage during overtime is 50% more than regular wage. This leads to an overtime cost factor of 1.5. Considering the case of *std_h* = 8, when there is no overtime, the number of work periods required to complete the task is 29.28. The labor cost can be determined from (29.28 periods)(8 h/periods)(total head count per shift)(regular wage per hour). Since the total head count per shift and regular wage remain the same for all cases, we will simply neglect these two and come up with total cost of 234.24, shown in Fig. 5(b) as 234.27 due to round-off numbers. In taking into account the overtime cost, say, fixed 1-h overtime when *std_h* = 8, the total cost will be based on regular work hours and overtime hours. This amounts to total cost (25.36 periods)(8 h/period) + (25 OT hours)(1.5 OT factor) = 240.4. Resulting total costs for all cases are shown on the right table in Fig. 5(b). Such result can be used to construct tradeoffs curve between overtime decisions, as shown in resulting work periods versus total cost. For example, in the case of *std_h* = 8 h, the tradeoff curve can be created as shown in Fig. 6.

As a general trend, one can expect inverse relationship between the two. The more overtime hours, the less number of work periods required and the higher total cost. This general trend is shown as trendline in Fig. 6. In this particular case, however, the 4-h overtime drops off from the trendline, representing a good place to be in, simultaneously reducing both work periods required and total cost. In fact, the total cost of fixed 4-h OT is about the same as that of fixed 1-h OT, but one can achieve 25.36–17.54 = 7.82 work period reduction. Constructing the tradeoff curve provides an intuitive tool to support overtime decisions well. As another example, Fig. 7 denotes the tradeoff curve when standard working hour is 16 h. There is no outlier in this case, the optimal overtime decision should be *fixed_OT* of 3 h.

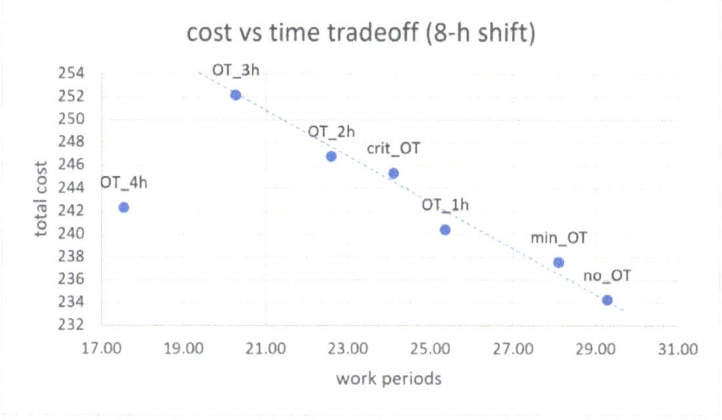

Fig. 6. Tradeoffs between completion time vs total cost for 8-h shift.

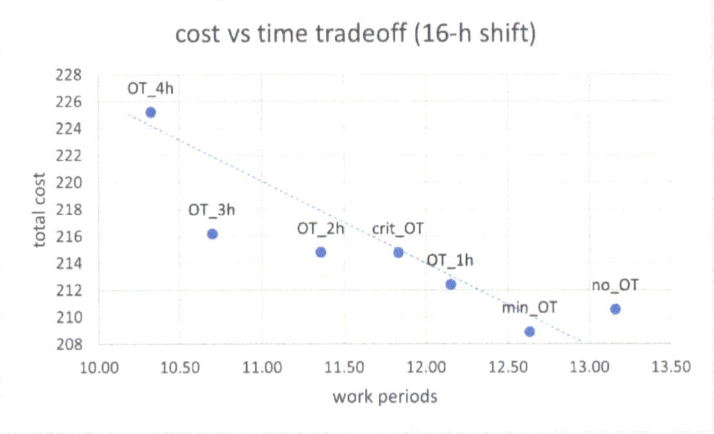

Fig. 7. Tradeoffs between completion time vs total cost for 16-hour shift.

5 Conclusion

In this research, a simulation tool is developed to adjust job schedule based on given standard working hours and overtime decision. The tool can be used to identify tradeoffs between adding overtime and speeding the work. A benchmark test case with the best-known solution, TA73, is used as a case study to demonstrate the impact of both standard working hours and overtime decision on makespan and work periods required. The tradeoff between the two can be shown visually and intuitively.

This research work is still preliminary. A number of aspects worth exploring include various overtime formats and the density of starting job schedules.

References

1. Jain, A.S., Meeran, S.: Deterministic job-shop scheduling: past, present and future. Eur. J. Oper. Res. **113**(2), 390–434 (1999). https://doi.org/10.1016/S0377-2217(98)00113-1
2. Shylo, O.V.: Job shop scheduling HomePage. https://optimizizer.com/TA.php. Accessed 31 Jan 2026
3. Taillard, E.: Benchmarks for basic scheduling problems. Eur. J. Oper. Res. **64**(2), 278–285 (1993). https://doi.org/10.1016/0377-2217(93)90182-M
4. Zeng, Q., Wang, M., Shen, L., Song, H.: Sequential scheduling method for FJSP with multi-objective under mixed work calendars. Processes **7**(12), 888 (2019). https://doi.org/10.3390/pr7120888
5. Google OR Tools. https://developers.google.com/optimization/scheduling/job_shop. Accessed 31 Jan 2026
6. IBM CP Optimizer HomePage. https://www.ibm.com/docs/en/icos/22.1.0?topic=cp-optimizer. Accessed 31 Jan 2026
7. Xiong, W., Li, Z.B., Hao, J.B.: Research on methods to solve key problems of dynamic job shop scheduling. Modul. Mach. Tool Autom. Manuf. Tech. **7**, 105–108, 112 (2008)
8. Goodwin, J.C., Elvers, D., Goodwin, J.S.: Overtime usage in a job shop environment. Omega **6**(6), 493–500 (1978). https://doi.org/10.1016/0305-0483(78)90051-8

9. Lagodimos, A.G., Mihiotis, A.N.: Overtime vs. regular shift planning decisions in packing shops. Int. J. Prod. Econ. **101**(2), 246–258 (2006). https://doi.org/10.1016/j.ijpe.2004.12.028
10. Artigues, C., Gendreau, M., Rousseau, L.M., Vergnaud, A.: Solving an integrated employee timetabling and job-shop scheduling problem via hybrid branch-and-bound. Comput. Oper. Res. **36**(8), 2330–2340 (2009). https://doi.org/10.1016/j.cor.2008.08.013
11. Thenarasu, M., Rameshkumar, K., Anbuudayasankar, S.P.: Milti-criteria decision making approach for minimizing makespan in a large-scale press-shop. Int. J. Ind. Eng. **26**(6) (2019)
12. Cloutier, S., Quimper, C.G.: Cumulative scheduling with calendars and overtime. In: 30th International Conference on Principles and Practice of Constraint Programming (CP 2024), Germany (2024). https://doi.org/10.4230/LIPIcs.CP.2024.7
13. Weise, T.: Job shop scheduling problem instances and results HomePage. https://github.com/thomasWeise/jsspInstancesAndResults. Accessed 31 Jan 2026

A Comparative Evaluation of Machine Learning and Deep Learning Models for Aspect Category Identification

Jantima Polpinij(✉), Vuttichai Vichianchai, Theeraya Utha, Anirut Chottanom, Jatuphum Juanchaiyaphum, Thananchai Khamket, and Chumsak Sibunruang

Faculty of Informatics, Mahasarakham University, Mahasarakham 44150, Thailand
{jantima.p,vuttichai.v,theeraya.u,anirut,jatuphum.j, thananchai.k,chumsak.s}@msu.ac.th

Abstract. Aspect category identification is a fundamental task of ABSA, which contributes to precisely knowing users' opinions rather than just sentiment polarity. Although aspect-based methods have been well investigated, direct comparisons between traditional machine learning and deep learning models in common experimental settings are still lacking. In this paper, it conducts a comparative study on representative machine learning (ML) and deep learning (DL) methods for aspect category extraction. Experimental results are reported on a benchmark dataset of 2,000 English hotel reviews downloaded from TripAdvisor, which have been manually labeled at the sentence level with respect to four aspect categories (i.e., staff service; cleanliness; value for money; and location convenience). This work adopts one unified experimental protocol with the same preprocessing, feature representation, and stratified cross-validation to cope with data imbalance. It also compares the classical classifiers using TF–IDF features with deep learning models that can make use of word embeddings and learn on sequences. In the experiment, it found that the deep learning model LSTM is better than traditional models in terms of RMSE and MAPE. The empirical results shed light on model behavior in practice and serve as a reference for relevant real applications to choose appropriate models for aspect category identification.

Keywords: Aspect Category Identification · Aspect-Based Sentiment Analysis · Text Classification · Machine Learning · Deep Learning

1 Background

Aspect-based sentiment analysis is an essential task in text mining and natural language processing [1, 2]. Beyond sentiment polarity, identifying sentence-level aspect categories—such as staff service, cleanliness, value for money, and location—enables a more fine-grained understanding of user opinions and supports better decision-making in service-oriented domains [3]. Given the importance of aspect category identification within ABSA, it has emerged as a critical classification problem in the literature [1].

S. Sodsee et al. (Eds.): IC2IT 2026, LNNS 5590, pp. 134–143, 2026.
https://doi.org/10.1007/978-3-032-28048-0_13

Previous research has investigated different methods for aspect-related tasks, such as classic machine learning models based on lexical and statistical features or deep learning architectures that can capture dependencies in context [4, 5]. Survey studies have shown that deep learning methods, including CNNs and LSTMs, dominate traditional classifiers for aspect-based tasks, as they can capture the semantic and sequential dependencies in a sentence [4–6]. Nevertheless, such conclusions are typically based on a diverse range of experimental set-ups, feature representations, or end-to-end models that bundle together several subtasks, which complicates direct comparison across model classes [1, 5].

The literature review reveals a significant gap in systematic comparative assessment [2]. Although substantial research has focused on proposing alternative model architectures or improving implicit aspect identification, relatively little attention has been paid to treating aspect category detection as an independent multi-class classification task and evaluating machine learning and deep learning approaches under identical experimental settings [7–9]. Consequently, it remains unclear whether observed performance differences stem from model architecture, feature representation, or variations in task formulation [2].

This leads to the research problem addressed in this study: How do traditional machine learning and deep learning models differ in their effectiveness for aspect category identification when evaluated within a unified experimental framework? Addressing this problem is essential for understanding the practical trade-offs between performance, complexity, and interpretability in aspect-oriented text classification [10].

Thus, the objective of this paper is to compare representative machine learning and deep learning models for aspect category identification in a comparative manner by adopting a unified dataset, feature representation, and evaluation protocol. Our research objective is to provide empirical guidance for informed model choice for aspect-level text analysis tasks, especially in low-resource settings where computational efficiency and robustness are important.

The contributions of this study are threefold. First, the task of aspect category identification is discussed as a multi-class text classification problem. Second, a consistent experimental setup is introduced for an unbiased comparison between classic machine learning classifiers and deep learning models. Third, the empirical results provide practical advice on the capabilities and weaknesses of various model families, which helps in understanding more easily how our models act toward aspect-oriented text classification.

The structure of this paper is organized as follows. Section 2 provides a description of the dataset used in this study. Section 3 outlines the research methodology. Section 4 presents and discusses the experimental results. Finally, Sect. 5 summarizes the conclusions.

2 Dataset Used

In this paper, we use a benchmark dataset provided by hotel reviews to compare machine learning and deep learning models in performing aspect category identification. The dataset was obtained from the TripAdvisor platform, where it includes 2,000 English hotel reviews that represent different hotel experiences in the hospitality domain. The neutral sentiment reviews (3-star ratings) were removed from the dataset to avoid ambiguous classification.

Each review was tokenized into sentences to facilitate finer-grained aspect analysis. In order to guarantee the quality and consistency of annotation, three linguistic specialists manually assigned each review sentence to the observed aspects among predefined aspect categories often adopted in hotel review analysis. Four dimensions in particular, comprising staff service, cleanliness, value for money, and location convenience, have been taken into account. These classes represent fundamental service aspects that are often found in customer reviews and commonly used in aspect-based sentiment analysis works.

Aspect category identification is posed as a multi-class classification task in this work, where each sentence is mapped to one of four aspect categories. While the original annotations make this distinction at the level of aspect experiences and expressions, we only used (explicit and implicit) aspect category labels in our study. This setup enables a fair and controlled comparison of different model families on the same data scenario.

The corpus is skewed in the distribution of aspect categories and expression types, as well as in the proportion between explicit and implicit aspect targets. Such inherent bias makes the task of aspect category identification harder for sentences that are more dependent on context rather than words. As a result, the dataset offers a realistic and tough benchmark to assess the capacity of various models for understanding semantic and contextual information beyond keywords in texts. The distribution of sentences among aspect categories and expression types is summarised in Table 1.

Table 1. Summary of the Aspect Category Dataset.

Aspect Category	Explicit	Implicit	Total
Staff Service	397	153	550
Cleanliness	385	165	550
Value for Money	300	150	450
Location Convenience	333	117	450
Total	1,415	585	2,000

All sentences of the review dataset were annotated manually by three linguistic experts following well-defined guidelines in order to guarantee their quality. Annotation inconsistencies were resolved through consensus to minimize ambiguity and improve labeling consistency. This expert-based strategy enforces the trustworthiness of the dataset for supervised aspect category detection.

The dataset has an imbalanced distribution over aspect categories and expression types, and this mirrors the naturally occurring user-generated hotel reviews. Unlike using synthesized resampling, it kept the original distribution to represent realistic conditions. Stratified 10-fold cross-validation was used to avoid the effect of imbalance during evaluation by maintaining similar class distributions in all folds. Several evaluation metrics, such as Accuracy and F1-score, were utilized to report unbiased model performance. This configuration allows for a reasonable comparison of machine learning with deep learning models in a realistic and challenging data setting.

3 Research Method

This section describes the employed methodology to compare ML and DL models for aspect category identification with a unified experimental setting. In order to make a fair comparison, we use the same dataset, label definition and evaluation protocol for all models and apply model-specific preprocessing and input representations.

3.1 Task Definition and Notation

Let $\mathcal{D} = \{(x_i, y_i)\}_{i=1}^{N}$ be the dataset about sentences, in which xi is one review sentence and $y_i \in \{1,2,3,4\}$ is the corresponding aspect category label: staff service, cleanliness, value for money, and location convenience. The aspect category identification can be cast as a problem of multi-class classification, which aims to learn a mapping $f(x_i) \rightarrow y_i$. The study follows the same aspect category labels for explicit and implicit sentences as presented in Sect. 2.

3.2 Overall Experimental Pipeline

The complete experimental pipeline consists of five consecutive steps in order to enable a balanced and systematic comparison between the models. Specifically, firstly, the reviews are divided into sentence-level instances, and each sentence of the instance is labeled with an aspect category as the ground truth. Then, model-specific text preprocessing steps are executed, differentiating those used for traditional machine learning models and those that must be applied for deep learning models. Next, the proper input representations are produced: for machine learning methods, these are TF–IDF vectors, and for deep learning techniques, they are preprocessed word sequences and a set of embedding-based inputs. In the fourth stage, stratified 10-fold cross-validation is used to train all models so as to ensure the stability and balance of class distribution among folds. Eventually, the models' performances in terms of Accuracy, Precision, Recall, and F1-measure are obtained, which makes a fair comparison of various model families under the same test protocol.

3.3 Text Preprocessing

The preprocessing is controlled to exclude possible confounding factors. The common pipeline units in both pipelines are just a few basic cleaning operations, and representation-oriented functions are different.

3.3.1 Preprocessing for Machine Learning Models (TF-IDF-Based)

Machine-learning classifiers (i.e., Support Vector Machines: SVM, Logistic Regression: LR, Multinomial Naive Bayes: MNB, Random Forest: RF) are trained on sparse lexical representations; therefore, the preprocessing step is dedicated to vocabulary normalization and noise reduction. Each preprocessing step can be described as follows [11, 12].

Step 1: Text normalization

All text is first pre-processed to achieve a unified and consistent textual representation by converting all characters into lowercase, removing punctuation marks and whitespace as well as non-alphabetic symbols.

Step 2: Sentence tokenization

This normalized text is further split into tokens (which are increasingly taken to be words) with a simple popular tokenizer as the one of NLTK toolkit.

Step 3: Stopword removal

The stop words as well as frequent function words (e.g., "the", "is", and "was") are excluded from the analysis to minimize dimensionality of features and highlight semantically informative terms.

Step 4: Lemmatization

Each token is then converted to its root or lemmatized form (e.g., "*cleaned*" and "*cleaning*" are both mapped to the base form, "*clean*") for reduction of data prevalence and less morphological variations.

Step 5: Feature construction

Finally, we build TF-IDF feature vectors of the processed tokens in which each term is weighted according to its frequency within a sentence and across the corpus. These TF–IDF representations are then regarded as feature vectors for the SVM, LR, MNB, and RF classifiers. The formula of TF-IDF is:

$$TF - IDF(t, x_i) = tf(t, x_i) \cdot \log(1 + \frac{N}{df(t)}) \quad (1)$$

The TF–IDF weight of term *t* for document x_i is measured based on term frequency and inverse document frequency. In particular, $tf(t, x_i)$ is the frequency of term *t* in document x_i and is a measure of its relevance in the document. The inverse document frequency part, $log(1 + N/df(t))$, quantifies how discriminative term *t* is to the documents in the corpus, where *N* is the total number of documents and $df(t)$ is the number of documents containing term *t*. The logarithmic scaling and addition are to prevent division by zero and smooth down factors from high occurrence terms.

3.3.2 Preprocessing for Deep Learning Models (Sequence-Based)

Deep learning models (i.e. CNN and LSTM) take in a sequence of words as input and gains from maintaining word order; hence preprocessing centres around sequence fidelity and controlled truncation/padding [13, 14].

Step 1: Text normalization

All input sentences are normalized to lowercase and then stripped of punctuation marks, as well as special symbols in our case, to standardize textual input.

Step 2: Tokenization for sequences

Each sentence is tokenized to form a sequence of tokens and retains the original word order to be used in sequence-based modeling.

Step 3: Vocabulary construction

A vocabulary *V* is built with only the training folds so as to avoid information leakage and each token is mapped to a unique integer index. It is note that Unlike TF-IDF pipelines, deep learning does not necessarily need to use stopword removal as embeddings and other sequence models are able to learn patterns locally within a sentence, thus requiring much less cleaning so we keep the sentences natural.

Step 4: Sequence length standardization

Maximum sequence length L is pre-defined, so the sentences which are less than L in length are padded and the ones exceeding the maximum sequence length are truncated to keep only first L tokens.

Step 5: Embedding input preparation

Every token is also mapped to a dense vector embedding learned by the Word2Vec Skip-gram model with dimension embedding d. As a result, each sentence xi is converted into an embedding matrix $X_i = \mathbb{R}^{L \times d}$, which constitutes the input of the deep learning models.

3.4 Feature Representation

In order to guarantee a fair comparison between the model families, we use two kinds of feature representation in this paper. Traditional machine learning approaches typically employ TF–IDF representations of term salience and discriminative lexical cues based on word frequency statistics. On the other hand, deep learning models use word embeddings which are trained to encode both semantic similarity as well as contextual meaning of words within a continuous vector space.

In addition to these regular representations, an optional weighted embedding representation is employed to derive a unified semantic feature for sentence level information. Formally, given a sentence x_i, we calculate its vector representation, denoted by v_i using weighted sum of word embedding vectors where the weight of each word is the TF–IDF score of that word in the input sentence such as [15, 16]:

$$v_i = \sum_{w \in x_i} TF - IDF(w, x_i) \cdot e_w \tag{2}$$

where e_w is the Word2Vec vector of word w. Such a construction involves both lexical saliency and semantical contribution, resulting in a cross-sentential one.

3.5 Model Training

All the models are trained by stratified 10-fold cross-validation to combat class imbalance problem and keep the original distribution of aspect categories. In each fold, nine subsets are used for training and one for testing. The hyperparameters are tuned on the training folds, thus avoiding information leakage and comparing all models under the same evaluation protocol.

Support Vector Machine (*SVM*) [17] - A linear kernel is used for SVM since it uses the much dimension of TF–IDF features. The parameter C is tuned by grid search on a predefined rang so as to balance between margin maximization and classification error.

Logistic Regression (*LR*) [18] - LR is optimized with $L2$ regularization, and the inverse of regularization strength C is fine-tuned to balance the model's capacity (complexity) and leverage overfitting.

Multinomial Naïve Bayes (*MNB*) [19] - The smoothing parameter α is optimized by MNB for zero-frequency problem solving and smooths the probability estimate, thereby increasing probability estimation's robustness, especially when using a sparse TF–IDF representation.

Random Forest (*RF*) [20] - To train the Random Forest classifier, parameters such as number of trees as 100, maximum tree depth as 10, and minimum samples per split as 5 are tuned for a trade-off between model bias-variance and generalization.

Convolutional Neural Network (*CNN*) [21] - The CNN models are trained on word embedding input and hyper-parameters, such as the number of filters, kernel size, dropout rate and learning rates, are tuned heuristically. Early stopping is used with validation loss as a measure to stop overfitting.

Long Short-Term Memory (*LSTM*) [22] - The LSTM model takes the input sequences one by one to learn long distance contextual dependencies. Hyperparameters tuning for deep learning models were done in the training folds to prevent information leakage. Hyperparameters such as hidden size, dropout and the batch size were set to 128, 0.5 and 64 respectively. Training was stopped at the maximum of 10 epochs with early stopping based on validation loss to prevent overfitting.

3.6 Evaluation Metrics

The performance of the models is measured by Accuracy, Precision, Recall and F1-Score with scores averaged over all folds. This one unified evaluation protocol ensures a fair and clear comparison between traditional machine learning and deep learning detectors [23].

4 Experimental Results

This section reports our experimental results of the comparison between traditional machine learning and deep learning-based aspect category identification. All the experiments were performed on the merged dataset as well as using the same preprocessing approach, feature representations, and performance evaluation described in Sects. 2 and 3, respectively. The goal is to study how models behave in the same setting and to determine performance differences due to model architecture rather than the experimental setup.

The average performance of all tested models using stratified 10-fold cross-validation is shown in Table 2. The obtained results are presented in terms of Accuracy, Precision, Recall, and F1-score, all of which satisfy the balanced evaluation scheme under the imbalanced condition.

From the results, it is evident that deep learning models achieve superior performance in comparison with conventional machine learning classifiers for all metrics. Overall, the performance of LSTM is superior to both CNN and traditional classifiers across all models.

The discrepancies in performance between the model families evaluated can be attributed to their representational capacity. Conventional machine learning classifiers are built on largely sparse TF–IDF representations, which encode lexical significance

Table 2. Performance Comparison of Machine Learning and Deep Learning Models for Aspect Category Identification.

Model	Accuracy	Precision	Recall	F1-score
SVM (Linear)	0.552	0.552	0.698	0.519
Multinomial Naïve Bayes	0.262	0.519	0.562	0.530
Logistic Regression	0.545	0.491	0.545	0.508
Random Forest	0.565	0.516	0.565	0.530
CNN	0.820	0.792	0.820	0.801
LSTM	0.882	0.902	0.906	0.904

but are not well-suited for contextual dependence. Therefore, these models achieve good performance when aspect categories are clearly shown by keywords but have limitations in dealing with implied aspects.

In contrast, deep learning models exploit dense word embeddings and sequence-dependent learning. The excellent performance of CNN indicates that both local n-gram patterns and semantic knowledge are informative in aspect discrimination. Instead, the LSTM model enhances the predictive power of implicit aspect cues distributed over a sentence by considering long-range contextual dependencies. This benefit is especially critical since the amounts of explicit and implicit aspect expressions in the dataset are imbalanced.

The rather mediocre performance of classical classifiers also indicates one important aspect, namely the issue related to an imbalanced dataset. Even though stratified cross-validation reduces evaluation bias, models without strong contextual modeling are still vulnerable to biased class distributions. In this respect, deep learning models (in particular the LSTM) exhibit more robust behavior; this further confirms their suitability for real-life aspect category extraction tasks.

5 Conclusion

This paper provided a comparative examination of ML and DL models for aspect category identification over the same dataset, with a common preprocessing pipeline and evaluation process. The experimental results demonstrate that conventional machine learning approaches relying on TF–IDF features have a poor ability to encode contextual and implicit aspect information, especially on imbalanced data. However, deep learning models using word embeddings and sequence modeling show much better performance than the above models, among which LSTM achieves the best results. Our results demonstrate that model architecture is critical in accurately detecting aspect categories when contextual dependencies are important. As further work, we can extend our approach by integrating transformer-based language models to more deeply exploit their capabilities in dealing with implicit aspect expressions and data imbalance while maintaining computational efficiency suitable for real-world deployment.

Acknowledgement. This work was financially supported by Mahasarakham University.

References

1. Zhang, W., Li, X., Deng, Y., Bing, L., Lam, W.: A survey on aspect-based sentiment analysis: tasks, methods, and challenges. IEEE Trans. Knowl. Data Eng. **35**(11), 11019–11038 (2022)
2. Trisna, K., Jie, H.: Deep learning approach for aspect-based sentiment classification: a comparative review. Appl. Artif. Intell. **36**(1) (2022)
3. Maroof, A., Wasi, S., Jami, S.I., Siddiqui, M.S.: Aspect-based sentiment analysis for service industry. IEEE Access **12**, 109702–109713 (2024)
4. Do, H.H., Prasad, P., Maag, A., Alsadoon, A.: Deep learning for aspect-based sentiment analysis: a comparative review. Expert Syst. Appl. **118**, 272–299 (2019)
5. Brauwers, G., Frasincar, F.: A survey on aspect-based sentiment classification. ACM Comput. Surv. **55**(4), 1–37 (2021)
6. Liu, H., Chatterjee, I., Zhou, M., Lu, X., Abusorrah, A.: Aspect-based sentiment analysis: a survey of deep learning methods. IEEE Trans. Comput. Soc. Syst. **7**(6), 1358–1375 (2020)
7. Busst, M.M.A., Anbananthen, K., Kannan, S., Kannan, R.: Extracting explicit and implicit aspects using deep learning. Emerg. Sci. J. **8**(1) (2024)
8. Elmobark, N.: Evaluating the trade-offs between machine learning and deep learning: a multi-dimensional analysis. J. Comput., Softw., Program **2**(1), 10–18 (2025)
9. Xue, W., Zhou, W., Li, T., Wang, Q.: MTNA: a neural multi-task model for aspect category classification and aspect term extraction on restaurant reviews. In: Proceedings of the International Joint Conference on Natural Language Processing (IJCNLP), pp. 1–13 (2017)
10. Wu, H., Wang, Z., Liu, M., Huang, J.: Multi-task learning based on question-answering style reviews for aspect category classification and aspect term extraction. In: Proceedings of the International Conference on Advanced Cloud and Big Data, pp. 1–8 (2019)
11. HaCohen-Kerner, Y., Miller, D., Yigal, Y.: The influence of preprocessing on text classification using a bag-of-words representation. PLoS One **15**(8), e0237406 (2020)
12. Pimpalkar, A., Retna Raj, R.R.: Influence of pre-processing strategies on the performance of ML classifiers exploiting TF–IDF and BoW features. Adv. Distrib. Comput. Artif. Intell. J. **9**(3), 25–38 (2020)
13. Sutskever, I., Vinyals, O., Le, Q.V.: Sequence to sequence learning with neural networks. In: Proceedings of the 28th International Conference on Neural Information Processing Systems (NeurIPS 2014), pp. 3104–3112 (2014)
14. Reddy, D.M., Reddy, S.: Effects of padding on LSTMs and CNNs (2019). arXiv preprint: arXiv:1903.07288
15. Gupta, S., et al.: Task-optimized word embeddings for text classification representations. Front. Appl. Math. Stat. **6** (2020)
16. Kenter, T., de Rijke, M.: Short text similarity with word embeddings. In: Proceedings of the 24th ACM International Conference on Information and Knowledge Management (CIKM 2015), pp. 1411–1420. ACM (2015)
17. Syarif, I., Prügel-Bennett, A., Wills, G.: SVM parameter optimization using grid search and genetic algorithm to improve classification performance. TELKOMNIKA **14**(4), 1502–1509 (2016)
18. El-Koka, A., Cha, K.-H., Kang, D.-K.: Regularization parameter tuning optimization approach in logistic regression. In: Proceedings of the International Conference on Advanced Communication Technology (ICACT 2013), pp. 1–5 (2013)
19. Juan-Císcar, A., Ney, H.: Reversing and smoothing the multinomial naive Bayes text classifier. In: Pattern Recognition in Information Systems, pp. 1–10 (2002)

20. Probst, P.: Hyperparameters, tuning and meta-learning for random forest and other machine learning algorithms. PhD thesis, University of Freiburg, Germany (2019)
21. Adamuthe, A., Jagtap, S.: Comparative study of convolutional neural network with word embedding technique for text classification. Int. J. Intell. Syst. Appl. **11**(6), 1–10 (2019)
22. Palangi, H., et al.: Semantic modelling with long short-term memory for information retrieval. arXiv preprint: arXiv:1412.6629 (2014)
23. Le Jeune, L., Goedemé, T., Mentens, N.: Machine learning for misuse-based network intrusion detection: overview, unified evaluation and feature choice comparison framework. IEEE Access **9**, 63995–64015 (2021)

Hybrid Transformer-Based Multimodal Fusion for Chest X-ray Report Generation

Van Van Chau[1], Tuan Minh Nguyen[1], and Phayung Meesad[2](✉)

[1] Faculty of Information Technology, Posts and Telecommunications Institute of Technology, 11 Nguyen Dinh Chieu, Ho Chi Minh City 700000, Vietnam
{vancv,minhtuan}@ptit.edu.vn

[2] Department of Information Technology and Management, King Mongkut's University of Technology North Bangkok, Bangsue 10800, Thailand
phayung.m@itd.kmutnb.ac.th, s6807011966511@kmutnb.ac.th

Abstract. Automated chest X-ray report generation has gained increasing attention as a tool to assist radiologists in high-volume clinical settings. Although recent advances in deep learning and Vision Transformer (ViTs) architectures improve visualtextual representation learning, challenges remain in balancing linguistic fluency, clinical accuracy, and robustness across heterogeneous pathological patterns [1]. This study presents a hybrid ViTs-based multimodal fusion framework that combines global visual encoding with transformer-based text generation to produce concise and clinically coherent reports. An optimization-aware decoding mechanism is introduced to regulate diagnostic consistency and suppress redundant or weakly supported statements [2]. Experiments on a publicly available chest X-ray dataset demonstrate consistent improvements over representative baselines across linguistic and clinical evaluation metrics, suggesting the framework's potential relevance for clinical decision support applications.

Keywords: Chest X-ray report generation · ViTs · Multimodal fusion · Clinical text generation · Medical image analysis

1 Introduction

Chest X-ray imaging remains a primary diagnostic modality in routine clinical practice because of its accessibility, low cost, and diagnostic utility across a wide spectrum of thoracic diseases, including pneumonia, tuberculosis, and lung tumors [1]. In high-volume settings, radiologists must interpret large numbers of examinations within limited timeframes, increasing the likelihood of reporting delays and variability across observers. These practical constraints have encouraged the development of automated systems capable of generating preliminary reports directly from imaging data. Early chest X-ray analysis methods were largely built upon convolutional neural networks (CNNs), which demonstrated

S. Sodsee et al. (Eds.): IC2IT 2026, LNNS 5590, pp. 144–153, 2026.
https://doi.org/10.1007/978-3-032-28048-0_14

reliable performance for disease classification and abnormality detection [3]. Nevertheless, CNN-based models are inherently constrained by their localized receptive fields, limiting their ability to capture long-range spatial relationships that are often relevant in diffuse or globally distributed radiographic patterns. Transformer architectures have recently provided an alternative paradigm. Vision transformers (ViTs) represent images as sequences of patches and employ self-attention to model global contextual interactions, thereby improving the representation of anatomical structures and subtle abnormalities [4]. In parallel, transformer encoderdecoder frameworks have become prominent in radiology report generation, strengthening alignment between visual features and textual outputs [5]. Despite these advances, transformer-based report generation models may still produce redundant descriptions, omit relevant findings, or introduce statements insufficiently supported by image evidence. Such issues raise concerns regarding clinical reliability. Multimodal learning strategies have therefore been explored to jointly model visual and textual representations, enabling more coherent reasoning across modalities [6]. However, achieving a balance between expressiveness, robustness, and computational efficiency remains challenging, particularly under limited annotations or heterogeneous disease distributions. In this study, we introduce a hybrid transformer-based multimodal fusion framework for chest X-ray report generation. The approach combines ViT-based visual encoding with a refinement mechanism designed to regulate redundancy and enhance clinical consistency. Rather than relying solely on end-to-end generation, the proposed design incorporates an optimization-aware stage that evaluates candidate outputs under multiple objectives. The contributions of this work are threefold: (i) a hybrid multimodal fusion architecture integrating global visual representations with transformer-based text generation; (ii) an optimization-guided refinement strategy to balance linguistic quality and clinical alignment; and (iii) empirical evaluation on a public chest X-ray dataset demonstrating consistent improvements over representative CNN and transformer baselines. The remainder of the paper is organized as follows. Section 2 reviews related literature, Sect. 3 details the proposed framework, Sect. 4 reports experimental results, Sect. 5 discusses findings, and Sect. 6 concludes the study.

2 Related Work

ViTs have emerged as a strong alternative to convolutional architectures in medical imaging by enabling global context modeling through self-attention mechanisms. Comprehensive surveys have shown that transformer-based models can outperform CNN-based approaches across classification, segmentation, and detection tasks when sufficient training data are available [7]. Subsequent studies further examined the scalability and robustness of transformers, highlighting their ability to generalize across imaging modalities and anatomical regions. In the specific context of chest X-ray analysis, recent work has demonstrated that deep learning models, including transformer-based architectures, can achieve diagnostic performance comparable to radiologists under certain clinical settings [8]. Extensions of transformer architectures to medical image segmentation have further confirmed their effectiveness in capturing structural and spatial patterns relevant to clinical interpretation [9]. Automated radiology report generation has evolved from early CNN–RNN pipelines toward transformer-based encoderdecoder frameworks. Early recurrent cascade models focused on image-to-text mappings but were limited by sequential dependencies and linguistic redundancy [10,11]. Transformer-based approaches introduced more flexible attention mechanisms and conditioning strategies, enabling improved alignment between visual features and generated textual descriptions. More recent methods incorporated keyword adaptation and frequency-aware modeling to enhance factual correctness and clinical relevance in generated reports [12]. Beyond architectural design, several studies have emphasized the importance of optimization and evaluation strategies for medical text generation. Surveys on transformer optimization have highlighted challenges related to efficiency and generalization in real-world deployments [13]. Hybrid approaches that combine evolutionary algorithms with machine learning have shown potential in refining objective functions and improving output quality [14]. Evaluation of generated radiology reports typically relies on both linguistic and clinical metrics, including BLEU [15], ROUGE [16], METEOR [17], and label-based clinical measures derived from expert annotations [18]. Evolutionary computation libraries such as DEAP have facilitated multi-objective optimization in text generation tasks [19], while large-scale datasets, including MIMIC-CXR, enable systematic benchmarking of report generation models [20]. Unlike existing works that apply optimization primarily for parameter tuning, our method integrates optimization directly into the report refinement stage, allowing explicit control over clinical consistency and redundancy in generated reports. In contrast to memory-driven or hierarchical cross-attention transformer frameworks such as R2Gen or M2Trans [21], the proposed approach explicitly separates draft generation from post-hoc refinement. Rather than embedding all control within the decoder architecture, the refinement stage evaluates multiple candidate reports under multi-objective constraints. This structural decoupling enables direct regulation of redundancy and clinical consistency, which is not inherently guaranteed by attention-based fusion mechanisms alone.

3 Our Proposed Method

This section presents a hybrid transformer-based multimodal fusion approach for chest X-ray report generation. The proposed method aims to improve clinical reliability while maintaining concise and fluent textual descriptions. Instead of relying solely on end-to-end text generation, the framework integrates global visual representation learning with an optimization-based refinement stage. As illustrated in Fig. 1, the proposed pipeline consists of three main components: (i) visual encoding using a vision transformer, (ii) neural report generation to produce an initial draft, and (iii) optimization-based refinement to enhance clinical alignment. The following subsections describe each component in detail.

3.1 Overall Workflow

Given an input chest X-ray image, the system first applies standard preprocessing, including resizing and normalization. The processed image is then encoded by a vision transformer to extract global visual representations that capture anatomical structures and potential abnormalities.

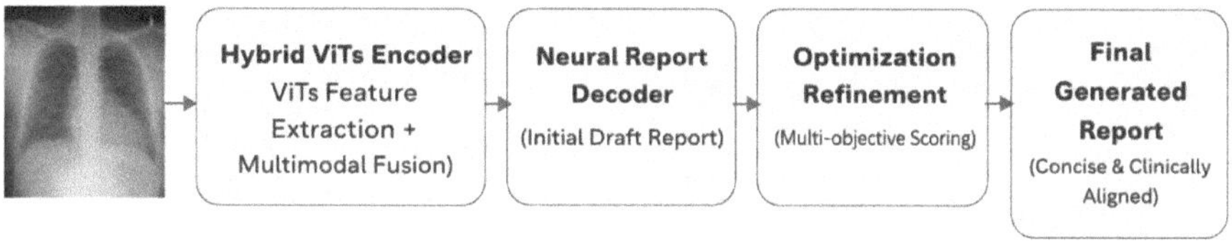

Fig. 1. Hybrid ViTs workflow for Report Generation

Based on these visual features, a neural decoder generates an initial draft report describing the observed findings. To further improve clinical consistency and reduce redundancy, an optimization layer refines the draft using multi-objective criteria. The final output is a concise and clinically accurate chest X-ray report.

3.2 Visual Encoding and Draft Report Generation

Let I denote an input chest X-ray image. The image is first divided into non-overlapping patches and projected into an embedding space before being processed by a vision transformer encoder. Through self-attention mechanisms, the encoder captures global dependencies across anatomical regions and produces a global visual representation v, corresponding to the classification token. Based on this visual representation, a transformer-based neural decoder generates an initial draft report S_0 in an auto-regressive manner. The decoder employs cross-attention to condition textual generation on visual features, ensuring that the generated text is grounded in image evidence. Although the draft report is generally fluent and informative, it may still contain redundant expressions or lack precise clinical alignment, motivating the refinement stage described next.

3.3 Optimization-Based Report Refinement

To improve clinical consistency and reduce redundancy, an optimization-based refinement strategy is applied to the initial draft report S_0. Starting from S_0, a population of candidate reports is generated through controlled variations. Each candidate is evaluated using a multi-objective fitness function that jointly considers linguistic similarity, clinical accuracy, and redundancy penalization:

$$F(S_i) = \alpha \cdot \mathrm{Sim}(S_i) + \beta \cdot \mathrm{Clin}(S_i) - \gamma \cdot \mathrm{Red}(S_i), \tag{1}$$

where α, β, and γ balance the contribution of each objective. Through iterative selection and population updates, the candidate with the highest fitness score is selected as the final optimized report. The complete optimization procedure is summarized in Algorithm 1.

Algorithm 1. Hybrid Transformer for Chest X-ray Report Generation

Require: Chest X-ray image I, patch size P, ViT encoder Φ.
Ensure: Optimized report S^*

$I' \leftarrow$ Preprocess(I) //resize, normalize
$v \leftarrow \Phi(I', P)$ //global visual feature (CLS token)
$S_0 \leftarrow \Psi(v)$ //generate initial draft report
$\mathcal{P} \leftarrow$ InitPopulation(S_0, N_{pop})
for $g = 1$ to G **do**
    for each $S_i \in \mathcal{P}$ **do**
        $\text{sim}_i \leftarrow$ SimScore(S_i) //e.g., BLEU/ROUGE/METEOR proxy
        $\text{clin}_i \leftarrow$ ClinScore(S_i) //e.g., CheXbert-F1 proxy
        $\text{red}_i \leftarrow$ RedPenalty(S_i)
        $F_i \leftarrow \alpha \cdot \text{sim}_i + \beta \cdot \text{clin}_i - \gamma \cdot \text{red}_i$
    end for
    $\mathcal{E} \leftarrow$ SelectTop$(\mathcal{P}, F, k)$
    $\mathcal{O} \leftarrow$ Crossover$(\mathcal{E})$
    $\mathcal{M} \leftarrow$ Mutate$(\mathcal{O})$
    $\mathcal{P} \leftarrow$ SurvivorSelect$(\mathcal{E} \cup \mathcal{M}, N_{\text{pop}})$
end for
$S^* \leftarrow \arg\max_{S_i \in \mathcal{P}} F_i$
return S^*

To quantitatively evaluate the effectiveness of the proposed framework for chest X-ray report generation, we compare our method against representative baselines, including a CNN–RNN architecture, a transformer-based model (ViTs), and an ImageNet100-pretrained transformer. Performance is assessed using standard language generation and clinical metrics, namely BLEU-4, ROUGE-L, METEOR, and CheXbert-F1. See Table 1.

4 Experiments and Results

4.1 Experimental Setup

Dataset. All experiments were conducted on the IU X-Ray dataset[1], which provides paired chest radiographs and corresponding radiology reports.

Data Split and Preprocessing. A study-level split was adopted, with 70% of the data used for training, 10% for validation, and 20% for testing. All images

[1] Chest X-Rays Indiana University (IU X-Ray), publicly available at https://www.kaggle.com/datasets/raddar/chest-xrays-indiana-university.

were resized to 224 × 224 and normalized. Text reports were tokenized and truncated or padded to a maximum length of 60 tokens, and a shared vocabulary was constructed from the training set.

Compared Models and Training Details. Four approaches were evaluated: CNN–RNN, ViTs, ImageNet100, and the proposed hybrid model. All models were trained using teacher forcing with cross-entropy loss and the AdamW optimizer (1×10^{-4}), batch size 16, with early stopping based on validation BLEU-4.

Evaluation Metrics. Linguistic quality was assessed using BLEU-1/2/3/4, ROUGE-L, and METEOR, while clinical correctness was evaluated using CheXbert-F1.

4.2 Quantitative Results

Table 1 summarizes the quantitative performance of all evaluated methods on the IU X-Ray dataset. Transformer-based approaches consistently outperform the CNN–RNN baseline, confirming the benefit of global contextual modeling for report generation. The ImageNet100 model further improves over the standard ViTs baseline, indicating the importance of stronger visual pretraining. The proposed method achieves the best performance across all metrics, demonstrating enhanced linguistic fluency and clinical alignment. Beyond aggregate metrics, we examined hallucinated findings, defined as clinically meaningful statements not supported by image evidence and identified through structured comparison with reference annotations. Compared with the base ViTs model, the refinement stage reduces their average occurrence per report by approximately 4050%, primarily by suppressing redundant or low-confidence phrases. This indicates that the gain extends beyond lexical similarity to stronger factual grounding.

Table 1. Performance on the IU X-Ray dataset for report generation.

Model	BLEU-4	ROUGE-L	METEOR	CheXbert-F1
CNN–RNN [22]	0.215	0.326	0.182	0.412
ViTs [23]	0.298	0.401	0.226	0.471
ImageNet100 [24]	0.312	0.414	0.234	0.489
Ours Model	**0.345**	**0.447**	**0.261**	**0.532**

5 Discussion and Evaluation

The qualitative examples in Fig. 2 demonstrate that the proposed framework produces structurally coherent reports that are largely aligned with the corresponding radiographic evidence. Most generated outputs accurately describe principal findings while maintaining concise textual structure, indicating that multimodal fusion effectively supports image-grounded report generation. Quantitative evaluation in Fig. 3 further confirms these observations, showing consistent improvements across both linguistic and clinical metrics compared with representative CNN- and transformer-based baselines. A closer inspection of generated outputs reveals that the refinement stage reduces unsupported or weakly grounded statements, thereby improving factual consistency beyond simple lexical overlap. Nevertheless, several failure cases highlight limited sensitivity to subtle or low-contrast abnormalities, which may reflect information attenuation during visual encoding. In addition, the refinement procedure introduces additional inference overhead, potentially restricting deployment in time-constrained clinical environments. Clinical correctness in this study is assessed using automated metrics such as CheXbert-F1, which function as structured proxies rather

Chest X-Ray	**Report Generation**
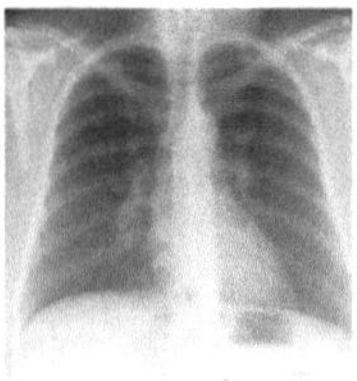	**Case 1 (No finding).** No active cardiopulmonary disease. The heart size and mediastinal contours are within normal limits. The lungs are clear bilaterally with no focal consolidation, pleural effusion, or pneumothorax.
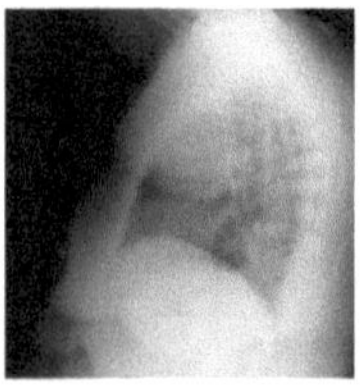	**Case 2 (Pneumonia suspected).** Patchy airspace opacity is observed in the right lower lung zone, suggestive of pneumonia. No pleural effusion or pneumothorax is detected. Heart size is within normal limits.
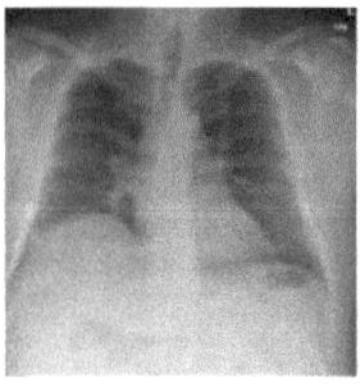	**Case 3 (Partial failure).** The lungs are grossly clear; however, the description misses subtle basilar opacity. No pneumothorax is identified. No acute bony abnormality is observed.
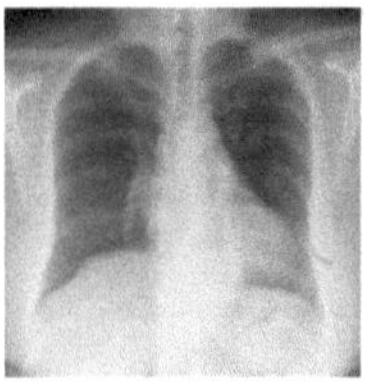	**Case 4 (Hiatal hernia).** No acute cardiopulmonary abnormality is detected. Heart size appears within normal limits. Overlying the cardiac silhouette suggests a hiatal hernia. No focal consolidation, pleural effusion, or pneumothorax is present.

Fig. 2. Qualitative cases of generated reports.

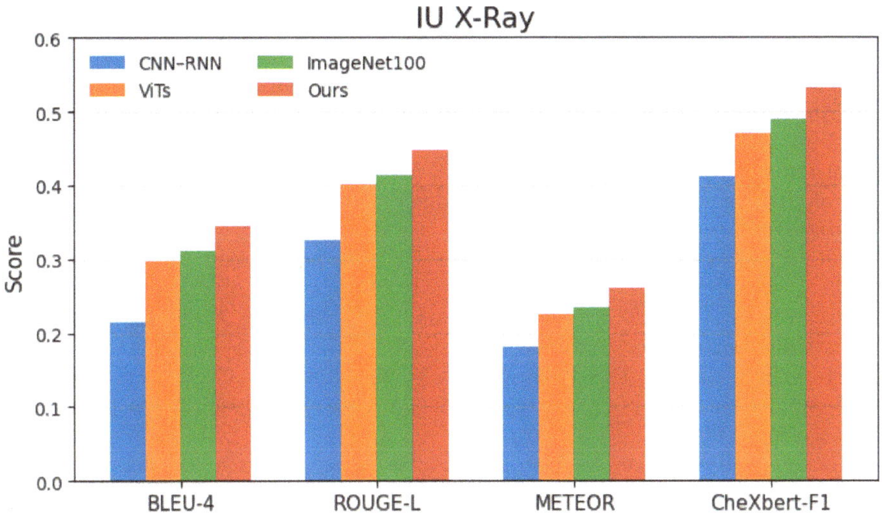

Fig. 3. Report generation performance on the IU X-Ray.

than substitutes for expert radiological evaluation. The absence of direct validation by board-certified radiologists therefore represents a limitation of the present work. Prospective assessment involving blinded expert review and safety-oriented evaluation protocols would be required prior to real-world clinical integration. Although a comprehensive sensitivity analysis was not exhaustively conducted, empirical variation of the refinement coefficients ($\pm 20\%$ around their default values) resulted in only minor fluctuations in evaluation metrics, suggesting that the optimization-based refinement mechanism maintains stable performance under moderate parameter perturbations.

6 Conclusion and Future Work

This paper presented a hybrid transformer-based multimodal fusion framework for chest X-ray report generation. By integrating global contextual modeling with an optimization-based refinement mechanism, the proposed approach achieves a favorable balance between linguistic quality, clinical consistency, and computational efficiency. Experimental results demonstrate consistent improvements over baseline models, supporting the applicability of the framework in practical clinical decision support scenarios. Future work will extend the evaluation to larger and more diverse datasets to assess generalization across imaging domains. In addition, applying the proposed framework to related tasks such as medical image analysis or image retrieval, as well as exploring more efficient refinement and attention mechanisms, may further enhance its scalability and real-time deployment potential. Broader validation across heterogeneous datasets and efficiency-oriented refinement strategies remain important directions for future investigation.

References

1. Irvin, J., et al.: CheXpert: A Large Chest Radiograph Dataset with Uncertainty Labels and Expert Comparison. Proc. AAAI Conf. Artif. Intell. **33**(01), 590–597 (2019). https://doi.org/10.1609/aaai.v33i01.3301590
2. Zhang, D., Ren, A., Liang, J., Liu, Q., Wang, H., Ma, Y.: Improving medical X-ray report generation by using knowledge graph. Appl. Sci. **12**(21), 11111 (2022). https://doi.org/10.3390/app122111111
3. Litjens, G., et al.: A survey on deep learning in medical image analysis. Med. Image Anal. **42**, 60–88 (2017). https://doi.org/10.1016/j.media.2017.07.005
4. Touvron, H., Cord, M., Douze, M., Massa, F., Sablayrolles, A., Jegou, H.: Training data-efficient image transformers & distillation through attention. In: Meila, M., Zhang, T. (eds.) Proceedings of the 38th International Conference on Machine Learning, vol. 139, pp. 10347–10357 (2021). https://proceedings.mlr.press/v139/touvron21a.html
5. Aksoy, N., Ravikumar, N., Frangi, A. F.: Radiology report generation using transformers conditioned with non-imaging data. In: Medical Imaging 2023: Imaging Informatics for Healthcare, Research, and Applications, 23 (2023). https://doi.org/10.1117/12.2653672
6. Shamshad, F., et al.: Transformers in medical imaging: A survey. Med. Image Anal. **88**, 102802 (2023). https://doi.org/10.1016/j.media.2023.102802
7. Azad, R., et al.: Advances in medical image analysis with vision Transformers: A comprehensive review. Med. Image Anal. **91**, 103000 (2024). https://doi.org/10.1016/j.media.2023.103000
8. Guo, L., Zhou, C., Xu, J., Huang, C., Yu, Y., Lu, G.: Deep learning for chest X-ray diagnosis: competition between radiologists with or without artificial intelligence assistance. J. Imaging Inf. Med. **37**(3), 922–934 (2024). https://doi.org/10.1007/s10278-024-00990-6
9. Pu, Q., Xi, Z., Yin, S., Zhao, Z., Zhao, L.: Advantages of transformer and its application for medical image segmentation: a survey. Biomed. Eng. Online **23**(1), 14 (2024). https://doi.org/10.1186/s12938-024-01212-4
10. Nguyen, L.D.V., Van Chau, V., Van Nguyen, S.: Face recognition based on deep learning and data augmentation. In: Dang, T.K., Kung, J., Chung, T.M. (eds.) Future Data and Security Engineering. Big Data, Security and Privacy, Smart City and Industry 4.0 Applications, pp. 560–573. Springer, Cham (2022)
11. Shin, H.-C., Roberts, K., Lu, L., Demner-Fushman, D., Yao, J., Summers, R. M.: Learning to read chest x-rays: recurrent neural cascade model for automated image annotation. In: 2016 IEEE Conference on Computer Vision and Pattern Recognition (CVPR), pp. 2497–2506 (2016). https://doi.org/10.1109/CVPR.2016.274
12. He, Z., Wong, A.N.N., Yoo, J.S.: Radiology report generation using automatic keyword adaptation, frequency-based multi-label classification and textto-text large language models. Comput. Biol. Med. **196**, 110625 (2025). https://doi.org/10.1016/j.compbiomed.2025.110625
13. Lin, T., Wang, Y., Liu, X., Qiu, X.: A survey of transformers. AI Open **3**, 111–132 (2022). https://doi.org/10.1016/j.aiopen.2022.10.001
14. Kumar, P.: A hybrid approach: combining genetic algorithms and machine learning for function optimization. J. Comput. Allied Intell. **3**(2), 56–65 (2025). https://doi.org/10.69996/jcai.2025012
15. Papineni, K., Roukos, S., Ward, T., Zhu, W.-J.: BLEU: A method for automatic evaluation of machine translation. In: Proceedings of the 40th Annual Meeting on

Association for Computational Linguistics - ACL '02, Vol. 311 (2001). https://doi.org/10.3115/1073083.1073135
16. Lin, C.-Y.: ROUGE: A package for automatic evaluation of summaries. In: Annual Meeting of the Association for Computational Linguistics (2004). https://api.semanticscholar.org/CorpusID:964287
17. Lavie, A., Agarwal, A.: Meteor: An automatic metric for MT evaluation with high levels of correlation with human judgments. In: Proceedings of the Second Workshop on Statistical Machine Translation - StatMT '07, pp. 228–231 (2007). https://doi.org/10.3115/1626355.1626389
18. Smit, A., Jain, S., Rajpurkar, P., Pareek, A., Ng, A., Lungren, M.: Combining automatic labelers and expert annotations for accurate radiology report labeling using BERT. In: Webber, B., Cohn, T., He, Y., Liu, Y. (eds.) Proceedings of the 2020 Conference on Empirical Methods in Natural Language Processing (EMNLP), pp. 1500–1519. Association for Computational Linguistics (2020). https://doi.org/10.18653/v1/2020.emnlp-main.117
19. Fortin, F.-A., Rainville, F.-M. D., Gardner, M.-A., Parizeau, M., Gagné, C.: DEAP: evolutionary algorithms made easy. J. Mach. Learn. Res. **13**, 2171–2175 (2012)
20. Johnson, A., et al.: MIMIC-CXR-JPG — chest radiographs with structured labels (version 2.1.0). PhysioNet. RRID:SCR 007345. (2024). https://doi.org/10.13026/jsn5-t979
21. Gholipour Picha, S., Al Chanti, D., Caplier, A.: Trust but verify: Imageaware evaluation of radiology report generators. Mach. Learn. Appl. **23**, 100851 (2026). https://doi.org/10.1016/j.mlwa.2026.100851
22. Jing, B., Xie, P., Xing, E.: On the automatic generation of medical imaging reports. In: Proceedings of the 56th Annual Meeting of the Association for Computational Linguistics (Volume 1: Long Papers), pp. 2577–2586 (2018). https://doi.org/10.18653/v1/P18-1240
23. Chen, Z., Song, Y., Chang, T.-H., Wan, X.: Generating radiology reports via memory-driven transformer. In: Proceedings of the 2020 Conference on Empirical Methods in Natural Language Processing (EMNLP), pp. 1439–1449 (2020). https://doi.org/10.18653/v1/2020.emnlp-main.112
24. Dosovitskiy, A., et al.: An Image is Worth 16x16 Words: Transformers for Image Recognition at Scale (No. arXiv:2010.11929) (2021). arXiv. https://doi.org/10.48550/arXiv.2010.11929

Predicting Corporate Dissolution in Thailand from Financial Statement Data Using Machine Learning Techniques

Pongtorn Panthai(✉) and Watchareewan Jitsakul

Faculty of Information Technology and Digital Innovation, King Mongkut's University of Technology North Bangkok, Bangkok, Thailand
s6707021856019@email.kmutnb.ac.th,
watchareewan.j@itd.kmutnb.ac.th

Abstract. This study aims to develop and compare the performance of machine learning models for predicting corporate bankruptcy in Thailand using financial statement data as the primary source. The sample consists of 13,218 legal entities, of which those that ceased operations in 2024 were classified as the event group. Financial statement data from 2022–2023 were used as independent variables for model construction. The machine learning algorithms employed in this study include Support Vector Machine (SVM), Light Gradient Boosting Machine (LightGBM), and Deep Neural Network (DNN). In addition, ensemble learning was applied using a voting-based approach to combine the predictions of multiple models. The data preparation process involved data cleaning, the construction of growth-related variables, and the transformation of categorical variables into suitable formats. Model performance was evaluated using K-fold cross-validation with Accuracy, Precision, Recall, and F1-score as evaluation metrics. The results indicate that the LightGBM model achieved the highest performance, with an accuracy of 94.10%, precision of 0.95, recall of 0.94, and F1-score of 0.94. Furthermore, when tested on unseen data not used during model training, the model maintained stable predictive performance.The findings suggest that machine learning models, particularly LightGBM, have strong potential for application as an effective decision-support tool for assessing corporate bankruptcy risk in Thailand.

Keywords: Bankruptcy · Machine Learning · Financial Statements

1 Introduction

In the current era of heightened global economic volatility and increasingly intense business competition, legal entities face substantial risks arising from both internal and external factors. Internal risks include weaknesses in management, financial instability, operational inefficiencies, and human resource challenges, while external risks stem from global economic fluctuations, technological disruption, shifts in consumer behavior, and changes in government policies. These combined pressures increase the likelihood of corporate failure and business cessation.

S. Sodsee et al. (Eds.): IC2IT 2026, LNNS 5590, pp. 154–165, 2026.
https://doi.org/10.1007/978-3-032-28048-0_15

In Thailand, business deregistration has risen significantly, with 23,380 firms ceasing operations in 2023—the highest level in the past 13 years—and risks remaining elevated in 2024 despite post-pandemic recovery. Corporate failure generates not only firm-level losses but also broader economic consequences for employees, creditors, investors, and the overall economy, particularly given the central role of SMEs in Thailand's economic structure.

A key challenge underlying this issue is the lack of an effective, data-driven framework for early identification of corporate failure risk using structured financial information. Although financial statements provide critical indicators of liquidity, leverage, and profitability, traditional analytical approaches may be insufficient to capture complex and nonlinear relationships among financial variables, limiting their predictive capability.

Recent advances in machine learning offer the ability to model complex patterns and improve predictive accuracy in large-scale datasets. However, empirical research applying and systematically comparing such techniques within the context of Thai legal entities especially using pre-failure financial statement data remains limited.

Therefore, this study seeks to address the problem of insufficient early risk detection by developing and systematically comparing machine learning–based models for predicting corporate failure among Thai legal entities. The objective is to enhance early warning capabilities and support more effective risk assessment and strategic decision-making for regulators, financial institutions, investors, and corporate managers.

2 Literature Reviews

2.1 Financial Statement Indicators for Business Failure Prediction

Correia [1] examined the role of accounting information in predicting corporate bankruptcy by conducting a comprehensive literature review and analyzing bankruptcy data of companies in the United States from 1980 to 2023. The study found that financial statement data and financial ratios continue to exhibit significant and stable predictive power for bankruptcy throughout the study period, despite changes in accounting standards and economic environments. At the same time, capital market data showed an increasing predictive contribution in more recent periods. However, models that integrate both accounting information and market-based data achieved the highest predictive performance. These findings indicate that accounting information remains a critical component and complements market data in assessing corporate failure risk.

Rejimon and Usha [2] investigated the use of financial statements and financial ratios to predict financial distress and bankruptcy, using state-owned enterprises in India as a case study from 2015 to 2019. The study applied several financial prediction models, including the Altman Z-Score, Logit Probability Model, and Fulmer Model, to evaluate liquidity, profitability, and financial stability. The results revealed that firms exhibited a high risk of bankruptcy during the early years of the study and showed clear signs of financial distress in the later period, particularly as indicated by liquidity indicators and long-term solvency measures. These findings suggest that financial ratios derived from financial statements can serve effectively as an early warning system for predicting corporate failure.

2.2 Machine Learning–Related Theories

2.2.1 Support Vector Machine (SVM)

SVM is a supervised machine learning algorithm widely used for classification and prediction tasks. Its core objective is to determine an optimal decision boundary that maximizes the margin between two classes. Given a training dataset (*xi*,*yi*), where *xi* is the input feature vector and $yi \in \{-1,1\}$ is the class label, SVM identifies the optimal separating hyperplane defined by the following equation:

$$w \cdot x + b = 0 \quad (1)$$

where w represents the weight vector, b denotes the bias term, and x is the input feature vector.

SVM constructs an optimal decision boundary by maximizing the margin from the support vectors, thereby enhancing generalization. To handle non-linear data, it projects inputs into a high-dimensional feature space. Due to its robustness, this study employs SVM as a benchmark model for predicting corporate failure [3].

2.2.2 Light Gradient Boosting Machine (LightGBM)

This study utilizes LightGBM, a highly efficient framework rooted in the Gradient Boosting Decision Tree (GBDT) paradigm. Instead of building independent estimators, the algorithm constructs an ensemble sequentially. By training each new tree to minimize the residual errors of its predecessors, the model progressively refines its predictive capability.

Boosting differs from bagging in that the learning process is sequential rather than parallel. In boosting, newly generated models place greater emphasis on instances incorrectly predicted by earlier models, allowing the algorithm to capture complex patterns and relationships in the data more effectively. LightGBM is specifically designed to achieve high computational efficiency by adopting a depth-wise tree growth strategy, which significantly reduces training time and memory consumption compared to traditional boosting algorithms.

Due to these characteristics, LightGBM is well-suited for handling large-scale datasets and high-dimensional feature spaces while maintaining strong predictive performance. As a result, it has been widely applied in classification and prediction tasks. In this study, LightGBM is employed as one of the primary algorithms for comparing the performance of corporate failure prediction models based on financial statement data [4].

2.2.3 Deep Neural Network (DNN)

DNN is a deep learning model derived from artificial neural networks, designed to capture complex relationships between input features and target outputs. Its defining characteristic is the presence of multiple hidden layers, which enables the model to learn highly nonlinear patterns more effectively than shallow networks (Schmidhuber, 2015). The term “deep” refers to the number of layers in the architecture rather than semantic depth.

The output of a DNN can be expressed as:

$$y(t) = \sum_{k=1}^{L} f(w_k + x_k(t)) + \varepsilon(t) \tag{2}$$

where w_k represents the weight matrix of each layer, $x_k(t)$ denotes the input sequence, f is the activation function, and $\varepsilon(t)$ is the error term.

DNNs achieve strong predictive performance on large-scale and high-dimensional datasets, although they require greater computational resources and training time. In this study, DNN is employed as one of the comparative models for predicting corporate failure based on financial statement data. [5].

2.3 Related Studies

Zhao, Ouenniche, and Smedt [6] proposed a bankruptcy prediction approach that integrates complex network analysis with traditional financial variables. By incorporating network-based metrics such as centrality measures and community structures, their results showed significant improvement in predictive accuracy, demonstrating the value of extending data sources beyond standalone financial statements.

Alanis, Chava, and Shah [7] conducted a comparative study of machine learning algorithms using long-term historical data and a forward-looking evaluation framework. Their findings indicated that gradient boosting models, including XGBoost and LightGBM, outperformed traditional statistical and baseline methods, supporting their role as strong benchmark models.

Hamdi, Mestiri, and Arbi [8] examined multiple techniques Logistic Regression, Decision Tree, SVM, Random Forest, and DNN for bankruptcy prediction in Tunisia. The DNN achieved the highest accuracy, highlighting the advantage of deep learning in capturing complex financial relationships.

Iparraguirre-Villanueva and Cabanillas-Carbonell [9] performed a comprehensive comparison of multiple machine learning models for business bankruptcy prediction using data from thousands of firms and a large number of financial variables. Their findings showed that ensemble models and well-tuned algorithms achieved higher accuracy and F1 Scores than baseline methods. This study highlights the importance of appropriate model selection and evaluation strategies in bankruptcy prediction tasks.

Kuizinienė et al. [10] conducted a systematic review of research on the application of artificial intelligence techniques for identifying corporate financial risk. By analyzing a wide range of studies that employed both machine learning and deep learning methods, the authors identified key factors influencing model performance, including handling imbalanced data, feature selection and dimensionality reduction, and the choice of appropriate evaluation metrics. This review provides strong methodological support for the framework adopted in the present study.

Overall, prior research particularly in international contexts demonstrates the strong predictive capability of advanced machine learning and deep learning models. However, empirical studies focusing on Thai legal entities, especially those utilizing pre-closure financial statement data under a unified evaluation framework, remain limited. This study addresses this gap by developing and systematically comparing corporate failure prediction models to support risk assessment and strategic decision-making in Thailand.

3 Research Methodology

This study focuses on predicting corporate failure among Thai legal entities using financial statement data and machine learning techniques. The objective is to develop predictive models capable of effectively assessing the risk of business cessation. The study's scope is divided into two main components: data-related and algorithm-related. The overall research framework is illustrated in Fig. 1.

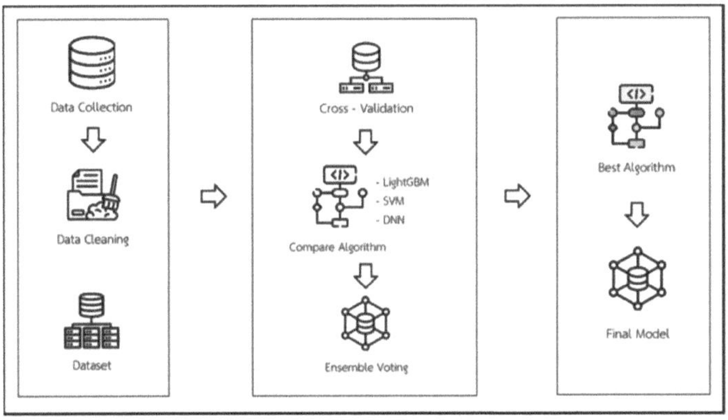

Fig. 1. Conceptual framework for research.

3.1 Collect Data

The dataset consists of financial information from 13,218 legal entities in Thailand and includes 32 variables covering corporate identification attributes, business characteristics, financial statement data for 2022–2023, financial ratios, auditor opinions, and operating status.

The data were imported and processed using Python and the pandas library. An initial exploratory data analysis was performed to examine observations, variables, data types, and missing values in order to understand the data structure prior to variable selection and transformation.

Target variable was constructed to classify operating status, where firms with "business cessation" were defined as the event group and others as continuing operations. The variable was encoded as binary for classification, with class 0 representing continuing operations and class 1 representing business cessation, as shown in Fig. 2.

Type of Corp.	TSIC	Region	Auditor's Opinion 2566	target
2	5	5	1	1
2	6	5	1	1
2	7	0	1	1
0	2	0	2	1
0	13	0	2	1
0	2	5	2	1
0	6	0	2	1

Fig. 2. Construction of the target variable.

3.2 Data Cleaning

Data cleaning is a critical step to ensure data quality and suitability for model development. In this study, variables were selected based on their theoretical relevance to financial risk and corporate failure and were categorized into three main groups: general corporate information variables, growth-related variables, and financial ratio variables. Growth-related features were constructed by comparing financial values for 2022 and 2023 to capture trends in assets, liabilities, revenue, and net profit. Categorical variables such as legal entity type, industry classification, business sector, and auditor opinion were encoded into numerical representations to ensure compatibility with machine learning models. The data-cleaning and feature-transformation process is illustrated in Fig. 3.

Firm Age	Total Asset 2566	Total Liabilities 2566	Total Revenue 2566	Net Profit (Loss) 2566	ROE (%) 2566	asset_ growth	debt_ growth	revenue_ growth	profit_ growth	debt_ratio _2566	profit_ margin_ 2566	Entity Type	TSIC	Region	Auditor's Opinion 2566	target
6	1.98E+08	1.9E+08	1.85E+08	914008	11.92	0.28	0.29	0.32	1.38	0.96	0.00	0	2	1	2	0
10	247695	201838.4	531968.1	123245.8	268.76	1.54	0.15	0.08	0.17	0.81	0.23	0	16	6	2	1
11	3011875	8000	0.07	-7999.93	-0.27	0.00	0.00	2.50	0.19	0.00	-114284.71	0	6	2	2	1
7	10755562	4541806	14575137	1704695	27.43	0.15	-0.07	0.45	36.15	0.42	0.12	0	5	3	2	0
14	26303266	40139281	2067266	-2087904	-15.09	-0.03	0.03	0.29	0.28	1.53	-1.01	0	11	0	2	0
7	2134844	2397032	11563197	347491.1	-132.54	-0.35	-0.39	0.45	1.39	1.12	0.03	0	8	6	0	0
12	4.22E+08	2.46E+08	1.08E+09	53973647	30.64	0.59	1.08	0.25	-0.11	0.58	0.05	0	2	2	2	1
10	1298822	5544	2836.35	-1663.65	-0.13	0.00	0.00	0.00	0.16	0.00	-0.59	2	6	3	1	1

Fig. 3. Construction of growth-related features and numerical encoding of categorical variables.

For missing values, entries that could not be converted into numerical form or resulted from division by zero were explicitly assigned as NaN, and no imputation was performed at this stage. The handling of missing values was deferred to the model development phase according to the characteristics of each algorithm.

To address class imbalance, the researcher balanced the dataset by selecting 5,000 firms with "business cessation" and 5,000 firms with "continuing operations" for model development, while the remaining observations were reserved as unseen data for out-of-sample evaluation.

3.3 Model Evaluation Using Cross-Validation

To ensure reliable and robust model evaluation, this study utilized 5-fold cross-validation. The performance metrics reported in the results section represent the average effectiveness of the model across these five folds.

3.4 Algorithm Performance Comparison

Machine learning models were developed and compared to evaluate their predictive performance. Three algorithms were considered in this study: SVM, LightGBM, and DNN. Hyperparameter tuning was performed for each model, with the selected parameter settings summarized in Table 1 for SVM, Table 2 for LightGBM, and Table 3 for DNN.

In addition to individual models, ensemble learning was applied to combine the predictions of multiple algorithms using a voting-based approach. This ensemble strategy was employed to examine whether integrating multiple classifiers could further enhance prediction performance.

Table 1. Hyperparameters used for tuning the SVM model.

Algorithm	Parameter	Value
Support Vector Machine	Kernel	poly, rbf, linear, sigmoid
	C	0.1, 0.2, 0.3, 0.4, 0.5, 0.6, 0.7, 0.8, 0.9, 1
	Gamma	0.1, 0.2, 0.3, 0.4, 0.5, 0.6, 0.7, 0.8, 0.9, 1

Table 1 Parameter settings considered for the SVM model were adjusted for performance testing. The kernel functions were poly, rbf, linear, and sigmoid. The regularization parameter C was set to 0.1, 0.2, 0.3, 0.4, 0.5, 0.6, 0.7, 0.8, 0.9, and 1, while the gamma parameter was set to 0.1, 0.2, 0.3, 0.4, 0.5, 0.6, 0.7, 0.8, 0.9, and 1.

Table 2. Parameter settings used for tuning the LightGBM model.

Algorithm	Parameter	Value
Light Gradient Boosting Machine	num_leaves	15, 31, 63
	min_child_samples	10, 20, 40
	max_depth	-1, 8
	learning_rate	0.05, 0.1
	n_estimators	100, 200, 300
	feature_fraction	0.8, 1.0
	reg_lambda	0.0, 0.1

Table 2 shows that the LightGBM model was tuned using several hyperparameter settings num_leaves was set to 15, 31, and 63; min_child_samples was set to 10, 20, and

40; max_depth was set to – 1 and 8; learning_rate was set to 0.05 and 0.1; n_estimators was set to 100, 200, and 300; feature_fraction was set to 0.8 and 1.0; and reg_lambda was set to 0.0 and 0.1.

Table 3. Hyperparameter settings used for tuning the DNN model.

Algorithm	Parameter	Value
Deep Neural Network	hidden_units	128, 64, 32
	dropout_rates	0.1, 0.1, 0.1; 0.3, 0.3, 0.2
	learning_rate	1e-3, 5e-4, 3e-4
	batch_size	256, 512
	epochs	100
	activation	ReLU
	use_bn	True
	l2	0.0, 1e-5, 1e-4

Table 3 shows that the DNN model was tuned using several hyperparameter settings.(hidden_units was set to 128, 64, and 32; dropout_rates were set to [0.1, 0.1, 0.1] and [0.3, 0.3, 0.2]; learning_rate was set to 1e-3, 5e-4, and 3e-4; batch_size was set to 256 and 512; epochs was fixed at 100; activation was ReLU; use_bn was enabled; and the l2 was set to 0.0, 1e-5, and 1e-4.

Table 4. Parameter settings considered for the ensemble learning technique.

Algorithm	**Parameter**
Ensemble	vote

Table 4 shows that the ensemble learning technique was configured using a single parameter, namely the vote parameter.

3.5 Final Model Development

After comparing the algorithms, the model with the best predictive performance based on the selected evaluation metrics was chosen as the final model. It was then trained on the entire cleaned and prepared dataset to learn comprehensive data patterns and ensure readiness for practical application.

The final model was subsequently tested on unseen data that had not been used during training or evaluation. This process assessed predictive capability under real-world.

conditions, and the results reflected the model's generalization ability for corporate failure prediction among legal entities in Thailand.

4 Results

This study presents the performance evaluation results of the developed corporate failure prediction models. The models were assessed using standard classification performance metrics, including accuracy (A), precision (P), recall (R), and F1-score (F), in order to comprehensively reflect predictive performance across different evaluation dimensions. The evaluation results are summarized in Tables 5, 6, 7 and 8.

Table 5. Results of the SVM model.

Parameter			Performance			
Kernels	Cs	gammas	A(%)	P	R	F
poly	**1**	**1**	**78.1**	**0.821**	**0.718**	**0.766**
poly	0.9	1	78.0	0.822	0.716	0.765
poly	0.8	1	77.8	0.821	0.713	0.763
poly	1	0.9	77.7	0.820	0.710	0.761
poly	0.9	0.9	77.6	0.821	0.707	0.760

Table 5 shows the best result of the SVM model with Kernel = "poly", C = "1", and gamma = "1". The result was an accuracy of 78.1%.

Table 6. Results of the LightGBM model.

Parameter							Performance			
num leaves	min_child samples	max_depth	learning rate	n-estimators	feature_fraction	reg_lambda	A(%)	P	R	F
15	**10**	**-1**	**0.05**	**300**	**0.8**	**0**	**94.1**	**0.957**	**0.924**	**0.940**
15	10	-1	0.05	300	0.8	0.1	94.1	0.957	0.923	0.940
15	20	8	0.1	100	0.8	0.1	94.1	0.957	0.923	0.940
15	40	8	0.05	300	0.8	0	94.0	0.956	0.923	0.939
15	40	8	0.1	100	0.8	0	94.0	0.956	0.923	0.939

Table 6 shows the best result of the LightGBM model with num_leaves = "15", min_child_samples = "10", max_depth = " − 1", learning_rate = "0.05", n_estimators = "300", feature_fraction = "0.8", and reg_lambda = "0". The result was 94.1% accuracy.

Table 7. Results of the DNN model.

Parameter					Preformance			
hidden_units	dropout_rates	learning_rate	batch_ size	12	A(%)	P	R	F
[128, 64]	**[0.1, 0.1]**	**0.0003**	**256**	**1.00E-05**	**87.1**	**0.843**	**0.918**	**0.878**
[128, 64]	[0.2, 0.2]	0.0003	256	1.00E-05	87.2	0.890	0.849	0.867
[128, 64, 32]	[0.3, 0.3, 0.2]	0.0005	256	0	86.0	0.882	0.826	0.850
[128, 64, 32]	[0.1, 0.1, 0.1]	0.0003	256	1.00E-05	85.1	0.865	0.837	0.847
[128, 64]	[0.2, 0.2]	0.0003	256	0	85.5	0.879	0.833	0.845
[64, 32]	[0.2, 0.2]	0.0003	256	1.00E-05	84.5	0.847	0.843	0.845

Table 7 shows the best result of the DNN model with hidden_units = "[128, 64]", dropout_rates = "[0.1, 0.1]", learning_rate = "0.003", batch_size = "256", and 12 = "0.00001". The result was 87.1% accuracy.

Table 8. Results of the ensemble model using the voting method.

Model	Performance			
	A(%)	P	R	F
SVM, LightGBM, DNN	93.1	0.931	0.931	0.931

Table 8 shows the best result of the ensemble model combining three algorithms, namely SVM, Light GBM, and DNN. The result was 93.1% accuracy.

Table 9. Performance evaluation results of the models using unseen data.

Model	Performance			
	A(%)	P	R	F
SVM	76.7	0.797	0.717	0.755
LightGBM	**92.9**	**0.933**	**0.925**	**0.929**
DNN	65.3	0.615	0.817	0.702
Ensemble Voting	92.71	0.928	0.927	0.927

Table 9 shows that, based on testing on unseen data comprising 3,186 observations, the LightGBM model achieved the best predictive performance, with an accuracy of 92.9%, precision of 0.933, recall of 0.925, and an F1-score of 0.929.

5 Conclusion

This study aimed to develop and compare the performance of machine learning models for predicting corporate failure among Thai legal entities using financial statement data. The dataset comprised financial information from 13,218 legal entities, where firms that ceased operations in 2024 were classified as the event group. Financial statement data from 2022–2023 were used as independent variables for model development.

Three machine learning algorithms were employed in this study: SVM, LightGBM, and DNN. Model performance was evaluated using standard classification metrics, including accuracy, precision, recall, and F1-score, under a unified evaluation framework. The experimental results indicate that the LightGBM model achieved the highest predictive performance among the models, with an accuracy of 94.1%, precision of 0.957, recall of 0.924, and F1-score of 0.940.

In addition, the models were further tested on unseen data not used during training or evaluation to assess their practical applicability. The results confirm that the LightGBM model consistently maintained high predictive performance, demonstrating strong generalization capability. These findings highlight the potential of the LightGBM model as an effective decision-support tool for risk assessment and strategic decision-making for Thai legal entities.

The superior performance of LightGBM over SVM and DNN is attributed to its robust handling of missing values and outliers inherent in financial ratios, as well as its leaf-wise growth strategy which effectively captures complex non-linear patterns in tabular data. In the context of the Thai economy, where SMEs face heightened post-pandemic volatility and liquidity constraints, this model serves as a critical early warning system. It enables regulators and financial institutions to proactively identify distressed firms and mitigate systemic economic risks associated with corporate dissolution.

References

1. Correia, M.: Accounting and corporate failure: the evolving role of accounting information in bankruptcy prediction, pp. 510–537 (2021). https://doi.org/10.1080/00014788.2025.2516273
2. Rejimon, A., Usha, M.: Financial distress analysis for the prediction of corporate bankruptcy – a case study of a public sector company in India, Salud, Ciencia y Tecnología - Serie de Conferencias **3**, 904 (2024). https://doi.org/10.56294/sctconf2024904
3. Tanma, K.: Personal Identification using global and local face analyses, pp. 11–13 (2023)
4. Wachwannakijkul, P.: Road travel time prediction between two coordinates in Bangkok using machine learning approaches, pp. 27–28 (2021)
5. Lertthaweedech, W.: Combining technical analysis and deep learning models for crude oil trading, pp. 4–6 (2021)
6. Zhao, J., Ouenniche J., Smedt J.D.: A complex network analysis approach to bankruptcy prediction using company relational information-based drivers. Knowl.-Based Syst. **300** (2024). www.elsevier.com/locate/knosys
7. Alanis, E., Chava S., Shah A.: Benchmarking machine learning models to predict corporate bankruptcy, pp. 23–24 (2022)
8. Hamdi, N., Mestiri, S., Arbi, A.: Artificial intelligence techniques for bankruptcy prediction of Tunisian companies: An application of machine learning and deep learning-based models. J. Risk Finan. Manag. **17**, 132 (2024). https://doi.org/10.3390/jrfm17040132

9. Iparraguirre-Villanueva, O., Cabanillas-Carbonell, M.: Predicting business bankruptcy: a comparative analysis with machine learning models. J. Open Innov. Technol. Mark. Complexity **10** (2024). https://doi.org/10.1016/j.joitmc.2024.100375
10. Kuizinienė, D., Krilavičius, T., Damaševičius, R., Maskeliūnas, R.: Systematic review of financial distress identification using artificial intelligence methods, pp. 3527–3571 (2023). https://doi.org/10.1080/08839514.2022.2138124

Prediction of Personal Loan Default Using Machine Learning

Nithikun Suntaranurak(✉) and Watchareewan Jitsakul

Faculty of Information Technology and Digital Innovation, King Mongkut's University of Technology North Bangkok, Bangkok, Thailand
s6707021856060@email.kmutnb.ac.th,
watchareewan.j@itd.kmutnb.ac.th

Abstract. This study aims to compare the performance of machine learning techniques for predicting delinquent personal loan borrowers. The dataset consists of 20,000 records with 11 features. All data were cleaned, transformed, and balanced using the SMOTE technique. The data cover personal loan information from January 2022 to July 2023, with repayment behavior tracked for an additional 18 months until February 2025. Borrowers were classified into two groups: non-delinquent and delinquent. Three algorithms were compared: Graph Convolutional Network (GCN), Extreme Gradient Boosting (XGBoost), and a Hybrid model. Model performance was evaluated using accuracy, precision, recall, and F1-score. The results show that the XGBoost model achieved the best performance. When tested on unseen data, it achieved 86.8% accuracy, with precision, recall, and F1-score all equal to 0.87.

Keywords: Graph Convolutional Network (GCN) · Extreme Gradient Boosting (XGBoost) · Personal Loan Default

1 Introduction

In 2023, Thailand's household debt increased to 91.3% of GDP, driven mainly by the expansion of consumer loans and housing loans. At the same time, non-performing loans (NPLs) in the consumer, real estate, and automotive sectors have continued to rise significantly since the pre-COVID-19 period. Thailand is also one of the countries with the highest household debt-to-GDP ratios in the world. This is reflected in the debt service coverage ratio (DSCR), which stands at 22.3%, well above the average of other major economies at 9.8%.

These factors, together with Thailand's policy interest rate at its highest level in the past 10 years and the combination of high household debt and relatively low repayment capacity, increase vulnerability to financial stability and the overall Thai economy in the future. This risk is particularly pronounced because a large proportion of Thailand's household debt comes from credit cards and personal loans.

Currently, the use of personal loans is increasing each year. The main reason people choose personal loans is to meet personal needs and cover essential daily expenses.

S. Sodsee et al. (Eds.): IC2IT 2026, LNNS 5590, pp. 166–178, 2026.
https://doi.org/10.1007/978-3-032-28048-0_16

However, lending by financial institutions always carries the risk that borrowers will become NPLs. As a result, many financial institutions must find ways to manage and mitigate this risk. In addition, stricter financial regulations on loan loss provisioning require institutions to set aside reserves under the TFRS 9 framework issued by the Bank of Thailand [1, 2], while also recognizing a loss of accrued interest income. These factors may increase the risk of operational disruption and liquidity shortages for financial institutions.

Therefore, most financial institutions have developed tools to reduce and prevent NPLs. One of these tools is a predictive model that forecasts loan delinquency based on various factors, such as customer demographics and repayment behavior. Such models help reduce the need for higher loan loss provisions. This research focuses on one of these key tools—delinquency prediction—by applying machine learning techniques to predict non-performing personal loan borrowers.

2 Literature Reviews

2.1 Graph Convolutional Network (GCN)

The basic idea of GCNs arises from the limitations of traditional models that treat data as tabular, assuming each row (sample) is independent of the others. In real-world applications, however, many datasets contain hidden relationships. For example, customers may share similar characteristics, belong to the same group, or exhibit similar behaviors. These relationships cannot be fully captured by individual features alone. GCNs are designed to learn from graph-structured data, consisting of nodes and edges [3, 4]. One key challenge in applying GCNs is that most tabular datasets lack an explicit graph structure. Therefore, a graph must first be constructed from the data. To address this issue, the k-Nearest Neighbors (kNN) approach is commonly used. In this step, each sample is connected to its k most similar neighbors based on feature similarity. The kNN-based graph construction strategy has been widely used in credit risk modeling to transform tabular data into relational structures [4, 5]. This similarity is usually measured using numerical distance metrics, such as Euclidean distance, after the features have been standardized to the same scale. In this framework, each data sample is treated as a node, and edges are created between a node and its nearest neighbors. This approach assumes that samples with similar features are more likely to share behaviors or states. When such a graph is fed into a GCN, the neighborhood aggregation process does not combine information from random neighbors; instead, it aggregates information from samples that are truly similar in terms of data characteristics. As a result, the learned embeddings capture both the intrinsic properties of each sample and the contextual information of its surrounding group.In other words, kNN acts as a bridge between tabular data and graph-based learning by transforming numerical similarity into a relational structure that GCNs can systematically learn from and utilize.

2.2 Extreme Gradient Boosting (XGBoost)

XGBoost is based on a fundamentally different idea. XGBoost belongs to the family of gradient boosting methods and has been widely applied in credit scoring and financial risk prediction [1, 2] due to its robustness and strong generalization ability [6, 7]. The core principle is to construct trees sequentially, with each new tree trained to correct the errors of the previous ones. The model starts with a rough prediction and gradually adds new trees that focus on samples mispredicted or poorly predicted in earlier iterations. This update process relies on gradients of the loss function, allowing the model to learn efficiently while maintaining strong control over overfitting. One of the main strengths of XGBoost is its ability to capture nonlinear relationships, threshold-based decision rules, and complex interactions among multiple features. For example, the effect of one feature may become significant only when another falls within a certain range, which decision trees handle very well. In addition, XGBoost is well-suited for tabular data, offers stable performance, and provides good interpretability through feature importance analysis.

2.3 Hybrid (XGBoost + GCN)

The hybrid approach combines the strengths of two algorithms. GCNs are used to learn the relational structure among samples, while XGBoost serves as a strong final decision model that is well-suited for tabular data and nonlinear interactions. Hybrid graph-based and boosting-based models have demonstrated superior performance in credit risk prediction tasks [5, 6, 8]. The key idea of the hybrid model is to use the GCN as a graph-based feature generator that produces new representations reflecting the context of each sample within a network, and then feed these representations into XGBoost, which is already known for its high predictive performance on tabular datasets. Conceptually, GCN focuses on "learning features from relationships," while XGBoost focuses on "classification or prediction" in an efficient and robust manner. In this framework, GCN acts as a structural feature extractor, and XGBoost serves as the predictive decision model [5, 6]. The first step of this approach is to transform tabular data into a graph. Each row in the dataset is treated as a node, representing one borrower. Edges between nodes are then defined to capture meaningful relationships. A kNN strategy is used to construct edges based on the proximity of scaled feature vectors (e.g., using StandardScaler), connecting borrowers who are numerically similar. In addition, group-based edges can be incorporated, such as linking borrowers who share the same status of residence, in order to form sub-communities that reflect specific population characteristics of interest. As a result, each node in the graph not only contains its own features but also gains contextual information from its neighbors, indicating which samples it is close to or grouped with.The core idea that makes this model truly hybrid lies in how the GCN outputs are used. The GCN-generated embeddings are combined with the original tabular features before being passed to XGBoost for prediction. This design is motivated by the strengths and limitations of each model. GCNs are highly effective at learning contextual information from relationships among samples, such as similarities or shared group membership. However, the neighborhood aggregation process in GCNs may smooth out or reduce the prominence of fine-grained numerical details, such as exact balances or subtle differences in key features. By retaining the original features alongside the GCN embeddings, the model preserves both perspectives simultaneously. In this combined representation, the GCN embeddings summarize the structural and relational context of each sample, while the original features maintain detailed, sample-specific information. XGBoost then integrates both sources of information, leveraging its ability to model nonlinear relationships and threshold-based decision rules to produce accurate predictions. In simple terms, the GCN creates a "shared representation" that captures contextual relationships in the data, while XGBoost "interprets" this representation together with the original features to deliver the final prediction as effectively as possible.

2.4 Related Work

Kamonchanok [1] developed a model to predict delinquent borrowers in the housing loan sector by comparing five algorithms: (1) SVM, (2) kNN, (3) XGBoost, (4) Random Forest, and (5) Deep Learning, along with an ensemble learning approach using majority voting. The study used a housing loan portfolio dataset covering the period from December 2018 to December 2019, comprising 28,418 records and 10 features. The data were divided into two parts: 19,896 records for model training and 8,522 records for model testing. Model performance was evaluated using accuracy, precision, recall, and F1-score. The results showed that the XGBoost algorithm achieved the best performance, with an accuracy of 86.84%, precision of 0.87, recall of 0.87, and F1-score of 0.87.

Kruawan and Sirisapp [2] studied the prediction of borrowers who are likely to default on bank loans using a credit card transaction dataset. The dataset consisted of 307,511 records with 122 features. The data were divided into two main groups: normal borrowers who did not default and abnormal borrowers who did. The researchers evaluated several machine learning techniques, including Logistic Regression, XGBoost, k-Nearest Neighbors (kNN), Random Forest, SVC, and Gradient Boosting. The results showed that the XGBoost model achieved the highest recall (0.97), indicating a strong ability to identify defaulting borrowers. However, the model achieved an accuracy of 0.37 and an F1-score of 0.51.

Sukharev et al. [3] aimed to develop a more accurate credit scoring model by incorporating intra-bank customer-to-customer money transfer data. The study demonstrated that customer connectivity and transaction relationships can significantly improve credit scoring performance compared to approaches that rely solely on individual customer attributes. The authors compared several algorithms, including GCN, GraphSAGE, Graph Attention Networks (GAT), Edge-Enhanced GNN (EGNN), Recurrent Neural Networks (RNN), and Light Gradient Boosting Machine (LightGBM). The results showed that the EWS-GCN model, a hybrid approach combining GCN and RNN, was the most effective at processing large-scale data. This model achieved an ROC score of 89.26%.

Shi et al. [4] attempted to develop a credit risk prediction model using graph representation learning combined with graph transformation techniques. The study utilized internal customer data from the German Credit, Lending Club, and Home Credit datasets, including features such as income, age, repayment history, number of accounts, debt ratio, and account duration. These variables were transformed into node features within a graph structure. The research compared a hybrid kNN–GNN model with traditional machine learning models, including Logistic Regression, Random Forest, and XGBoost. The results showed that the hybrid kNN–GNN model outperformed Logistic Regression and Random Forest, achieving 88.9% accuracy, 85.7% F1-score, and 91.3% AUC.

Wang et al. [5] designed a hybrid modeling framework that combines multiple algorithms to handle user data with both graph-structured relationships and temporal characteristics. The study used real-world data from Alipay and was developed for practical deployment in the Alipay application for borrower risk screening. The proposed framework compared several models, including XGBoost, GraphSAGE, LSTM, GAT, GCN–LSTM, ST-GCN, and TemGNN, which represents a hybrid ST-GCN–LSTM model. The experimental results showed that TemGNN achieved the best performance, with an accuracy of 84.3%, an F1-score of 85.6%, and an AUC of 87.9%.

Liu, Zhang, and Fan [6] conducted a study on credit risk prediction to estimate the probability of default for loan applicants using machine learning and big data techniques. Their goal was to develop an effective credit risk prediction model by comparing XGBoost, a GNN (ForgeNet), and a hybrid model combining both approaches to identify the best-performing method. The study used real-world loan data from Lending Club covering the period from 2007 to 2016. The results showed that the hybrid model achieved the best average performance, with an accuracy of 87.52%, an F1-score of 93.13%, and a G-mean of 85.59%.

Albanesi and Vamossy [7] developed models to predict consumer default risk using deep learning techniques, based on data from the housing crisis period between 2007 and 2009. The study compared several approaches, including Logistic Regression, Decision Trees, XGBoost, DNN, and a hybrid DNN-Gradient Boosting model. The results showed that the hybrid DNN–XGBoost model achieved the best performance across all time periods, with accuracy consistently above 86%, an AUC score exceeding 92%, and a Gini coefficient of approximately 86%.

Sun et al. [8] proposed a new approach for credit risk prediction using GCN to evaluate the creditworthiness of loan applicants. The study applied big data and artificial intelligence (AI) techniques, starting with transforming borrower data into a graph structure, where borrowers and their relationships were represented as nodes and edges, respectively. The proposed approach compared several models, including GCN, SVM, Decision Tree, backpropagation-based neural networks, and SGCNs, which represent a hybrid GCNN model. The experimental results showed that the SGCNs model achieved the best performance, with an average accuracy of 87.52%, an average F1-score of 93.13%, and an average G-mean of 85.59%.

3 Research Methodology

This study aims to compare the performance of different algorithms for predicting delinquent personal loan borrowers. The conceptual framework of the study is illustrated in Fig. 4.

3.1 Collect Data

The data used in this study consist of records of personal loan repayment behavior. The dataset includes information such as loan purpose, loan amount, outstanding balance, age, occupation, income, expenses, residential status, loan term, marital status, education level, type of residence, length of employment, length of employment at the current job,

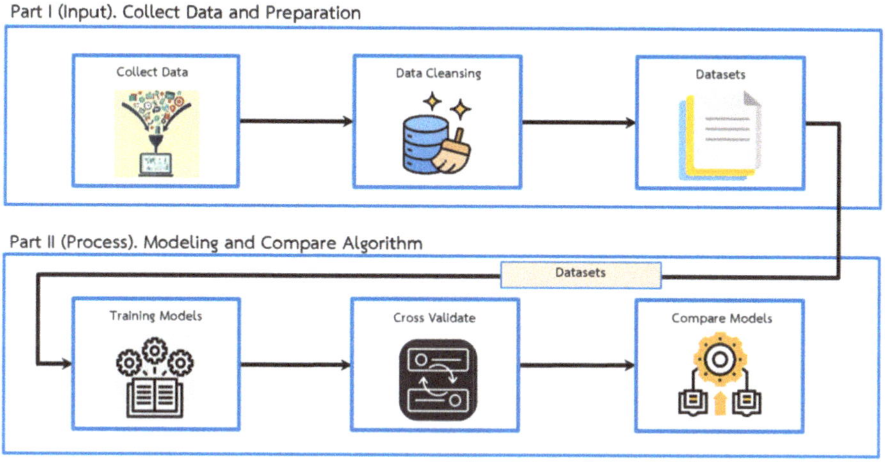

Fig. 4. Conceptual framework for research.

and business ownership. The dataset contains 50,000 records. An 18-month performance window was defined to classify accounts into two groups: non-delinquent customers (Good) and delinquent customers (Bad).

3.2 Data Cleaning and Transformation

The data were cleaned by removing records with missing or incomplete information. After the cleaning process, a total of 13,596 complete records remained, consisting of 10,000 non-delinquent customers and 3,596 delinquent customers. From the original variables, 11 relevant features were selected: loan purpose, loan amount, outstanding balance, age, occupation, income, loan term, marital status, education level, length of employment, and business ownership. Next, the data were converted to numerical values to facilitate model development. Categorical variables were encoded into numerical representations to ensure compatibility with the machine learning algorithms used in this study (Table 1).

Table 1. An example of transforming data into numerical representations.

Loan Amount	Value
1	1000–10000
2	10001–15000
3	15001–20000
4	20001–25000
5	25001–50000
6	50001–100000

(continued)

Table 1. (*continued*)

Loan Amount	Value
7	> 100000

3.3 Data Balancing

After data cleaning and transformation, the dataset remained highly imbalanced, with non-delinquent customers forming the majority class. To address this issue, the Synthetic Minority Over-sampling Technique (SMOTE) was applied to handle class imbalance. SMOTE generates synthetic samples by interpolating between existing minority-class observations, resulting in a more realistic distribution of delinquent customer data and improving the models' learning capabilities. In this study, the number of delinquent customer records (3,596) was increased to match the size of the non-delinquent customer group (10,000). As a result, 6,404 synthetic delinquent records were generated.

3.4 Data Balancing

3.4.1 Model Development Using the GCN Algorithm

The GCN model development process begins with graph construction, where numerical similarities between samples are used to define relationships via the kNN approach. After the graph is constructed, each row of the dataset is treated as a node, while edges represent the contextual relationships between samples. Predictions are primarily based on node embeddings that capture the graph's structural information. The GCN operation can be expressed as follows

$$\hat{y}_i = \text{softmax}\left(\mathbf{W}_{out}\, \sigma\left(\tilde{\mathbf{D}}^{-\frac{1}{2}}\tilde{\mathbf{A}}\tilde{\mathbf{D}}^{-\frac{1}{2}}\mathbf{X}\mathbf{W}\right)_i\right) \quad (1)$$

$\mathbf{X}$ is the initial feature matrix of all nodes (e.g., income, age, outstanding balance).

$\mathbf{A}$ is the adjacency matrix of the graph, representing connections between nodes.

$\tilde{\mathbf{A}} = \mathbf{A} + \mathbf{I}$ is the adjacency matrix with added self-loops.

$\tilde{\mathbf{D}}$ is the degree matrix of $\tilde{A}$, used to normalize the influence of neighboring nodes.

$\tilde{\mathbf{D}}^{-\frac{1}{2}}\tilde{\mathbf{A}}\tilde{\mathbf{D}}^{-\frac{1}{2}}$ is the graph convolution operator that balances information aggregation from neighbors.

$\mathbf{W}$ is the learnable weight matrix of the GCN layer.

$\sigma(\cdot)$ is a nonlinear activation function, such as ReLU.

$\mathbf{W}_{out}$ represents the weight matrix of the classifier layer.

softmax$(\cdot)$ converts the output into class probability distributions.

Table 2. Parameters used for tuning the GCN algorithm.

Parameter	Value
F	3, 5, 7, 9, 11, 13
dropout	0.1, 0.2, 0.5
Epochs	60, 100, 150
Hidden dim	32, 64, 128

From Table 2, the researchers tested 162 parameter settings. They set K to [3, 5, 7, 9, 11, 13], dropout to [0.01, 0.02, 0.05], epochs to [60, 100, 150], and hidden dimensions to [32, 64, 128].

3.4.2 Model Development Using the XGBoost Algorithm

XGBoost builds a model by stacking many decision trees. Each tree doesn't learn everything from scratch, but instead is trained to fix the mistakes made by the previous trees. This way, the learning gradually improves. During training, the model evaluates prediction quality using loss functions or metrics to avoid overfitting. Once training is complete, XGBoost outputs probabilities for the target classes, allowing you to adjust the threshold based on the context. Altogether, this makes XGBoost a great fit for tabular data with non-linear relationships and many feature interactions, as expressed below.

$$\hat{y}_i = \sigma\left(\sum_{m=1}^{M} f_m(\mathbf{x}_i)\right), \qquad f_m \in \mathcal{F}_{tree} \tag{2}$$

$\mathbf{x}_i$ is the feature vector of sample i

$f_m(\cdot)$ the m-th decision tree.

M is the total number of trees in the model.

$\mathcal{F}_{tree}$ is the set of functions that represent decision trees.

$\sum_{m=1}^{M} f_m(\mathbf{x}_i)$ means combining the outputs from all trees (boosting).

$\sigma(\cdot)$ is the sigmoid function, which converts a logit into a probability.

Table 3. Parameters used for tuning the XGBoost algorithm.

Parameter	Value
n_estimators	200, 400, 600, 800
learning_rate	0.03, 0.05, 0.1
max_depth	4, 6, 8, 10
min_child_weight	1, 2, 4
subsample	0.1, 0.2, 0.5, 1
colsample_bytree	0.01, 0.02, 0.05, 1

(continued)

Table 3. *(continued)*

Parameter	Value
reg_alpha	0, 0.1
reg_lambda	0, 1
gamma	0, 0.1

From Table 3, the researchers tested 18,432 parameter combinations. It set n_estimators (trees) to [200, 400, 600, 800], learning_rate to [0.03, 0.05, 0.1], max_depth to [4, 6, 8, 10], min_child_weight to [1, 2, 4], subsample to [0.1, 0.2, 0.5, 1], colsample_bytree to [0.01, 0.02, 0.05, 1], reg_alpha to [0, 0.1], reg_lambda to [0, 1], and gamma to [0, 0.1].

3.4.3 Model Development Using Hybrid Algorithm

The idea behind the hybrid approach is to split the work: GCN captures structural context, while XGBoost handles predictive decision-making. In practice, the hybrid process takes GCN embeddings and combines them with the original features before feeding the resulting features to XGBoost. This reflects the idea that graph embeddings provide the 'big picture context,' while raw features still give important 'numeric details.' By merging both, the final model doesn't lose information from either side. In this setup, GCN acts as a structural feature extractor, and XGBoost serves as the predictive decision model. It systematically combines the strengths of deep learning on graphs with tree-based models on tabular data. The parameters set in the hybrid model are the best values selected beyond those from GCN and XGBoost individually. The core concept is that GCN builds structural representations, and XGBoost uses those representations to learn complex prediction rules, as shown in the equation.

$$\hat{y}_i = \sigma\left(\sum_{m=1}^{M} f_m\left(\left[\mathbf{z}_i \,\|\, \mathbf{x}_i\right]\right)\right), \qquad \mathbf{z}_i = \mathrm{GCN}(\mathbf{x}_i, \mathcal{G}) \tag{3}$$

$\mathbf{z}_i$ is the embedding of node i from a GCN captures both its own features and the context from its neighbors

$\mathcal{G}$ is the graph structure (nodes + edges) used in the GCN.

$\mathbf{x}_i$ is the original feature of sample i (usually standardized).

$[\mathbf{z}_i \,\|\, \mathbf{x}_i]$ means concatenating the GCN embedding with the original feature,

$f_m(\cdot)$, M, $\sigma(\cdot)$ have the same meaning as in XGBoost only.

3.5 Model Evaluation

To evaluate the model's performance, we use standard classification metrics: Accuracy, Precision, Recall, and F1-score. These metrics are chosen to provide a well-rounded view of the classification's quality. Here's what they mean:

Accuracy measures the model's overall correctness. It's calculated as the proportion of correctly predicted samples compared to the total number of samples, as shown in the formula below.

$$\text{Accuracy} = \frac{TP + TN}{TP + TN + FP + FN} \tag{4}$$

TP (True Positive) = number of positive samples correctly predicted as positive.

TN (True Negative) = number of negative samples correctly predicted as negative.

FP (False Positive) = number of negative samples incorrectly predicted as positive.

FN (False Negative): The number of positive samples incorrectly predicted as negative.

Precision measures how reliable the model's positive predictions are. It's basically the proportion of correctly predicted positives compared to all the samples predicted as positive, as shown in the formula.

$$\text{Precision} = \frac{TP}{TP + FP} \tag{5}$$

Recall (also called Sensitivity) shows how well the model can actually detect the true positive samples. It's calculated as the proportion of positive samples correctly predicted among all positive samples, as shown in the formula.

$$\text{Recall} = \frac{TP}{TP + FN} \tag{6}$$

F1-score is the harmonic mean of Precision and Recall. It helps balance the evaluation by accounting for both types of errors—False Positives and False Negatives—simultaneously, as shown in the formula.

$$\text{F1-score} = \frac{2 \times \text{Precision} \times \text{Recall}}{\text{Precision} + \text{Recall}} \tag{7}$$

4 Experimental Results

The model's performance is evaluated using the following metrics: Accuracy, Precision, P-Recall and F1-Score (Table 4).

Table 4. The result of the GCN model.

Paramiter				Performance			
k	D-out	hidden	epochs	A	R	F	P
3	**0.1**	**128**	**150**	**0.78105**	**0.7492**	**0.7808**	**0.8002**
3	0.2	128	150	0.7796	0.7447	0.7793	0.8007
3	0.5	128	150	0.7757	0.7375	0.7754	0.7985
3	0.1	128	100	0.7726	0.742	0.7724	0.7905
3	0.1	64	150	0.7724	0.7374	0.7721	0.7929

(continued)

Table 4. (*continued*)

Paramiter				Performance			
k	D-out	hidden	epochs	A	R	F	P
3	0.2	128	100	0.7714	0.7389	0.7712	0.7903
3	0.2	64	150	0.77015	0.7289	0.7698	0.7944
3	0.5	128	100	0.76965	0.733	0.7693	0.7910
3	0.5	64	150	0.76915	0.7231	0.7687	0.7964
3	0.1	32	150	0.7679	0.7292	0.7675	0.7904

With K = 3, Dropout = 0.1, Hidden dimension = 128, and 150 epochs, the model achieved its best performance, reaching an accuracy of 78.11% (Tables 5 and 6).

Table 5. The result of the XGBoost model.

	M.1	M.2	M.3	M.4	M.5	M.6	M.7	M.8	M.9	M.10
colsample	**1**	1	1	1	0.5	0.5	0.5	0.5	0.5	0.5
gamma	**0**	0	0	0	0.1	0	0	0.1	0.1	0
learning	**0.1**	0.1	0.1	0.1	0.05	0.05	0.05	0.05	0.05	0.05
max_dept	**10**	10	8	8	6	6	6	6	6	6
min_child	**1**	1	1	1	1	1	1	1	1	1
n_estimate	**800**	600	800	600	800	800	800	800	800	800
alpha	**0**	0	0	0	0	0.1	0	0.1	0.1	0.1
lambda	**0**	0	0	0	0	0	0	0	1	1
subsampl	**1**	1	1	1	0.5	0.5	0.5	0.5	0.5	0.5
A	**0.9244**	0.9243	0.9171	0.911	0.8933	0.8929	0.8929	0.8925	0.8865	0.8859
P	**0.9279**	0.9276	0.9202	0.9137	0.8941	0.8937	0.8936	0.8933	0.887	0.8865
R	**0.9244**	0.9243	0.9171	0.911	0.8933	0.8929	0.8929	0.8925	0.8865	0.8859
F	**0.9242**	0.9242	0.9169	0.9109	0.8932	0.8928	0.8928	0.8924	0.8864	0.8859

The parameters set as n_estimators (trees) = 800, learning_rate = 0.1, max_depth = 10, min_child_weight = 1, subsample = 1, colsample_bytree = 1, reg_alpha = 0, reg_lambda = 0, and gamma = 0, the model achieved its best performance with an accuracy of 92.44%.

Table 6. The result of the hybrid model.

Model	Performance			
	A(%)	P	R	F
XGBoost + GCN	85.95	0.8735	0.8614	0.8498

The parameters from GCN set as K = 3, Dropout = 0.1, Hidden dimension = 128, and Epochs = 150, and the parameters from XGBoost set as n_estimators (trees) = 800, learning_rate = 0.1, max_depth = 10, min_child_weight = 1, subsample = 1, colsample_bytree = 1, reg_alpha = 0, reg_lambda = 0, and gamma = 0, the model achieved its best performance with an accuracy of 85.95%.

Table 7. Model performance results on unseen data.

	Performance			
Parameter	A(%)	P	R	F
GCN	49.20	0.240	0.4576	0.4839
XGBoost	**86.80**	**0.868**	**0.8676**	**0.872**
GCN + XGBoost	81.20	0.924	0.8309	0.7549

From Table 7., when testing all three models on the same unseen dataset of 500 samples, the XGBoost algorithm achieved the best performance, with an accuracy of 86.80%, an average precision of 0.87, an average recall of 0.87, and an average F1-score of 0.87.

5 Conclusion

This study presents a machine learning approach to predicting personal loan default. The dataset consists of 20,000 records collected between January 2022 and July 2023, with repayment behavior tracked for an additional 18 months until February 2025. Each record contains 11 features: loan purpose, loan amount, outstanding balance, age, occupation, income, loan term, marital status, education level, work experience, and business ownership. Borrowers were categorized into two groups: normal borrowers and default borrowers. Three algorithms were compared GCN, XGBoost and GCN + XGBoost model. Data preprocessing involved cleaning missing and incomplete records, transforming features into numerical values, and applying SMOTE to address class imbalance. Model evaluation was conducted using Stratified K-5 Fold Cross-Validation, with performance measured by Accuracy, Precision, Recall, and F1-score. The experimental results show that XGBoost achieved the best performance, with an accuracy of 92.44%, a precision of 0.93, a recall of 0.92, and an F1-score of 0.92. Further testing on an unseen dataset of 500 samples confirmed XGBoost as the most effective algorithm, yielding an accuracy of 86.80%, and average precision, recall, and F1-score of 0.87.

References

1. Rungrueang, K.: Forecasting mortgage loan default using data mining techniques Master of Science in Information Systems for Management, Department of Information Technology Management, Faculty of Information Technology, King Mongkut's University of Technology North Bangkok (2022)

2. Netprana, K., Laohakiat, S.: Risk analysis of credit card default. Master of Science in Data Science, Faculty of Science, Srinakharinwirot University (2023)
3. Sukharev, I., et al.: EWS-GCN: Edge Weight-Shared Graph Convolutional Network for Transactional Banking Data. arXiv preprint arXiv: 2009.14588. (2020)
4. Shi, Y., et al.: Improved credit risk prediction based on an integrated graph representation learning approach with graph transformation. Eur. J. Oper. Res. **315**(2), 786–801 (2024)
5. Wang, D., et al.: Temporal-aware GNN for credit risk prediction. In: Proceedings of the SIAM International Conference on Data Mining (SDM), pp.702–710 (2021)
6. Liu, J., Zhang, S., Fan, H.: A two-stage hybrid credit risk prediction model based on XGBoost and graph-based DNN. Proc. Expert Syst. Appl. **195**, 116624 (2022)
7. Albanesi, S., Vamossy, D.: Predicting Consumer Default: A Deep Learning Approach of a proceedings paper.: NBER, Working Paper No. 26165, pp. 1–72. National Bureau of Economic Research 1050 Massachusetts Avenue Cambridge (2019)
8. Sun, M., et al.: Applying Hybrid GNN to Strengthen Credit Risk Analysis. arXiv preprint arXiv: 2410.04283 (2024)

Forecasting Hydrological Inflows of Xayaburi Dam Lao People's Democratic Republic Using Deep Learning Techniques

Chalermchad Homat(✉) and Watchareewan Jitsakul

Faculty of Information Technology and Digital Innovation, King Mongkut's University of Technology North Bangkok, Bangkok, Thailand
s6707021857201@email.kmutnb.ac.th,
watchareewan.j@itd.kmutnb.ac.th

Abstract. This study focuses on forecasting hydrological inflows to the Xayaburi Dam in the Lao People's Democratic Republic using deep learning techniques. Daily hydrological and meteorological data collected from 1 January 2020 to 31 December 2024, comprising 1,827 records, were utilized. The dataset was divided into training data covering the period from 2020 to 2023 and testing data for the year 2024. Three deep learning models, namely Recurrent Neural Network (RNN), Long Short-Term Memory (LSTM), and Gated Recurrent Unit (GRU) were developed and evaluated. Model performance was assessed using five statistical metrics: Mean Squared Error (MSE), Root Mean Squared Error (RMSE), Mean Absolute Error (MAE), Mean Absolute Deviation (MAD), and Mean Absolute Percentage Error (MAPE). The experimental results indicate that the LSTM model outperformed the RNN and GRU models, achieving the lowest error values with an MSE of 0.3448, RMSE of 0.5872, MAE of 0.1981, MAD of 0.0608, and MAPE of 4.3236. Consequently, the LSTM model was selected as the optimal model and applied to forecast hydrological inflows to the Xayaburi Dam for the subsequent 365 days, from 1 January 2025 to 31 December 2025. The findings demonstrate that LSTM is effective in capturing temporal dependencies in hydrological time-series data and can provide valuable support for dam operation planning and water resource management.

Keywords: Hydrological inflow forecasting · Deep learning · Long Short-Term Memory (LSTM) · Recurrent Neural Network (RNN) · Gated Recurrent Unit (GRU)

1 Introduction

Dam-based water resource management is a critical component of national water security and sustainable development. Dams function as a strategic hydraulic infrastructure for regulating, storing, and distributing water in support of multiple operational objectives, including hydropower generation, agricultural irrigation, flood mitigation, and domestic water supply. Effective reservoir operation requires optimized allocation strategies to

S. Sodsee et al. (Eds.): IC2IT 2026, LNNS 5590, pp. 179–190, 2026.
https://doi.org/10.1007/978-3-032-28048-0_17

balance competing demands under variable hydrological conditions. However, operational decisions aimed at reducing flood risk may conflict with storage objectives for energy production, agricultural use, and municipal consumption. Excessive pre-release of reservoir water in anticipation of extreme precipitation events can result in water scarcity during subsequent dry periods, whereas sufficient discharge may cause reservoir levels to exceed design storage capacity, leading to uncontrolled spillway releases and sudden downstream flooding. Consequently, reservoir operation policies and decision-making processes significantly influence the performance and effectiveness of agencies responsible for dam operation and flood control.

In the context of forecasting hydrological inflows for water resources management, various algorithms have been proposed to analyze hydrological data. For instance, Recurrent Neural Networks (RNN) have been applied to forecast monthly streamflow in the Zarrine River basin [1] Gated Recurrent Units (GRU) have been used to develop flood forecasting models for the city of Ashland, Tennessee [2] and Transformer-based Long Short-Term Memory (LSTM) models have been employed to predict rainfall in mountainous regions of the northwestern Himalayas [3].

Based on the problems and related studies discussed above, this research recognizes the importance of forecasting hydrological inflows to the Xayaburi Dam in the Lao People's Democratic Republic, which is a key factor for effective reservoir water management. Accurate inflow forecasting is essential to prevent flooding and water scarcity in the region. Therefore, this study aims to develop a highly accurate hydrological inflow forecasting model to serve as a decision-support tool for future water management operations of the Xayaburi Dam. Ultimately, the proposed approach is expected to promote sustainable water resource utilization and maximize benefits for local communities and the country.

2 Literature Reviews

This study aims to investigate appropriate deep learning techniques for forecasting hydrological inflows to the Xayaburi Dam in the Lao People's Democratic Republic. To achieve this objective, relevant data and prior studies on hydrological inflow forecasting and deep learning applications are reviewed. The details are presented as follows:

2.1 Recurrent Neural Networks (RNN)

Recurrent Neural Networks (RNNs) are neural network models designed to process sequential data by maintaining an internal state that captures temporal dependencies. At each time step, the hidden state is updated based on the current input and the previous hidden state [4]. The formulation of an RNN is shown in Eqs. (1) and (2).

The hidden state update of an RNN is defined as:

$$h_t = f(Wx_t + Uh_{t-1} + b) \tag{1}$$

The output of the RNN at time step t is given by:

$$y_t = g(Vh_t + c) \tag{2}$$

where x_t is the input vector at time step t, h_t denotes the hidden state representing the internal memory of the network, y_t is the output at time step t, W, U, and V are weight matrices, b and c are bias terms, and $f(\cdot)$ and $g(\cdot)$ denote activation functions.

2.2 Long Short-Term Memory (LSTM)

Long Short-Term Memory (LSTM) networks incorporate a memory cell and gating mechanisms that regulate the flow of information over time. This structure enables effective retention, updating, and control of relevant temporal information across long sequences [5]. The formulation of an LSTM is shown in Eqs. (3) and (4).

The cell state update of an LSTM is defined as:

$$c_t = f_t \cdot (c_{t-1} + i_t \cdot \widetilde{c_t}) \tag{3}$$

The hidden state update of an LSTM is defined as:

$$h_t = o_t \cdot \tanh c_t \tag{4}$$

where c_t denotes the cell state, h_t denotes the hidden state, f_t, i_t, and o_t represent the forget, input, and output gates, respectively, $\tilde{c}_t$ is the candidate cell state, and $\bullet$ indicates element-wise multiplication.

2.3 Gated Recurrent Units (GRU)

Gated Recurrent Unit (GRU) networks employ gating mechanisms to regulate the flow of information over time, enabling effective modeling of long-term dependencies in sequential data. The architecture of a GRU utilizes a single hidden state representation and a set of gates to control information retention and update [6]. The formulation of a GRU is shown in Eqs. (5) to (8).

The update gate of a GRU is defined as:

$$z_t = \sigma(W_z x_t + U_z h_{t-1} + b_z) \tag{5}$$

The reset gate of a GRU is defined as:

$$r_t = \sigma(W_r x_t + U_r h_{t-1} + b_r) \tag{6}$$

The candidate hidden state of a GRU is defined as:

$$\tilde{h}_t = \tanh(W_t x_t + U_h(r_t \bullet h_{t-1}) + b_h) \tag{7}$$

The hidden state update of a GRU is defined as:

$$h_t = (1 - z_t) \bullet h_{t-1} + Z_t \bullet \tilde{h}_t \tag{8}$$

where x_t denotes the input vector at time step t, h_t denotes the hidden state, z_t and r_t represent the update gate and reset gate, respectively, $\tilde{h}_t$ is the candidate hidden state, $\sigma(\cdot)$ and $\tanh(\cdot)$ denote the sigmoid and hyperbolic tangent functions, respectively, and $\cdot$ indicates element-wise multiplication.

2.4 Related Work

Nakhaei et al. [1] presented a study on monthly streamflow forecasting in the Zarrine River basin, a major watershed feeding Lake Urmia in Iran. The study compared the forecasting performance of RNN and LSTM models using data collected from 1979 to 2014. Five model structures were developed with input time lags of up to four months. The results indicated that the RNN model achieved the highest forecasting accuracy, with RMSE = 22.8, R2 = 0.84, and NSE = 0.80.

Fordjour and Kalyanapu [2] presented a study on flood forecasting for the city of Ashland, Tennessee, using a GRU-based model. The study employed half-hourly water level data from ten gauging stations in the Lower Cumberland–Sycamore watershed. The results indicated that the GRU model achieved high forecasting accuracy, with NSE = 0.982, RMSE = 0.064 ft, MAE = 0.052 ft, and R^2 = 0.984.

Wani et al. [3] presented a study on rainfall forecasting in mountainous regions of the northwestern Himalayas. The study compared the forecasting performance of ANN, SVR, RF, KNN, LSTM, Bidirectional LSTM, Deep LSTM, RNN, GRU, ARIMA, and TBATS models using daily meteorological data collected from 1980 to 2020 at six stations located at different elevations. Model performance was evaluated using RMSE and MAE. The results indicated that the Bidirectional LSTM and LSTM models achieved the highest forecasting accuracy at several stations, particularly at station L1, where the LSTM model obtained RMSE = 38.28 and MAE = 22.27.

Shi et al. [7] presented a study on time-series forecasting of mine water inflow using an LSTM–Transformer model. The study compared the forecasting performance of LSTM, CNN, Transformer, LSTM–Transformer, and CNN–LSTM models using water inflow data collected from the Baotailong mine in Heilongjiang Province. The dataset was divided into training and testing subsets. The results indicated that the LSTM–Transformer model achieved the highest forecasting accuracy, with MAE = 11.2 and RMSE = 13.7.

Ameri et al. [8] presented a study on PM2.5 concentration forecasting using a CEEMDAN-based decomposition approach combined with SVM and LSTM models. The study compared the forecasting performance of SVM, LSTM, CEEMDAN–SVM, CEEMDAN–LSTM, and CEEMDAN–SVM–LSTM models using daily PM2.5 data collected in Kaohsiung City, Taiwan, from 2019 to 2021. The results indicated that for 1-day-, 3-day-, and 7-day-ahead PM2.5 forecasting, the CEEMDAN–SVM–LSTM model achieved the highest accuracy, with MAE = 1.858, MSE = 7.2449, RMSE = 2.6682, and NSE = 0.9169.

Jittraphorn [9] presented a study on the development of a rainfall forecasting model for drought analysis in the Nan River basin. The study compared the forecasting performance of LSTM and ARIMA models using monthly rainfall data collected from 15 rain gauge stations in the Nan River basin during the period from 1954 to 2021. The dataset was divided into a training period (1954–2004) and a testing period (2005–2021). Model performance was evaluated using RMSE, NSE, and the correlation coefficient (R). The results indicated that the LSTM model outperformed the ARIMA model at 10 out of 15 stations, achieving lower RMSE and higher NSE values.

Thitiwat [10] presented a study on the development of a rainfall forecasting model for drought analysis in the Mun River basin. The study compared the forecasting performance of LSTM and ARIMA models using monthly rainfall data from 27 rain gauge stations during the period from 1991 to 2020, along with drought records obtained from the Department of Disaster Prevention and Mitigation for the period 2011–2020. Drought conditions were analyzed using the Standardized Precipitation Index (SPI) at three time scales, namely SPI3, SPI6, and SPI12. Model performance was evaluated using RMSE, NSE, and the correlation coefficient (R). The results indicated that the LSTM model achieved superior performance at 21 out of 27 stations (77.78%), compared to 22.22% for the ARIMA model. Furthermore, the LSTM model was applied to perform a five-year-ahead drought forecast for the period from 2016 to 2020, and the predicted results were compared with calculated SPI values and observed drought events. The average forecasting accuracy over the five-year period was 76.30% and 54.07%, respectively.

3 Methodology

3.1 Research Framework

For hydrological inflow forecasting at the Xayaburi Dam in the Lao People's Democratic Republic, this study adopts a research framework, as illustrated in Fig. 1.

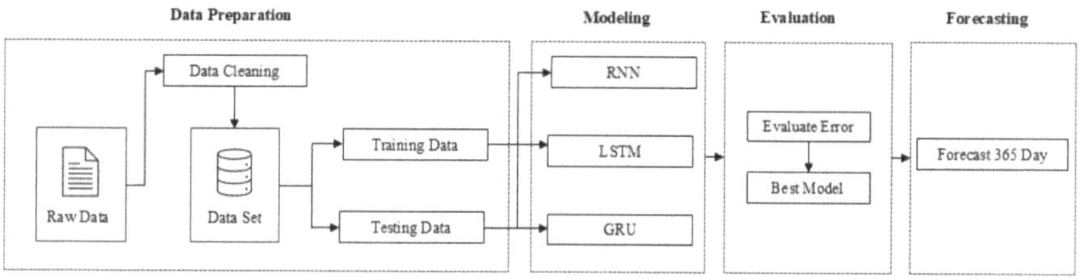

Fig. 1. Research framework.

3.2 Data Preparation

The data preparation process consisted of the following steps:

1. Hydrological inflow data of the Xayaburi Dam in the Lao People's Democratic Republic were collected for the period from 1 January 2020 to 31 December 2024. The dataset comprised daily records totaling 1,827 observations and included five attributes: inflow, outflow, rainfall, air temperature, and reservoir water level.
2. Data cleaning was performed to ensure data quality, including checking data completeness and handling missing values in the example dataset presented in Table 1.

Table 1. Example dataset

Date	inflow (m^3/s)	inflow (10^3 m^3/s)
01–01-20	1,981.49	1.98
02–01-20	2,088.59	2.09
03–01-20	1,995.79	2.00

3.3 Modeling

Based on the data preparation process, forecasting models were developed using three deep learning algorithms: RNN, LSTM, and GRU. The model development process consisted of the following steps:

1. The parameters of the RNN, LSTM, and GRU algorithms were defined and tuned to achieve the lowest possible prediction error. The parameters used for testing each algorithm are presented in Table 2.

Table 2. Parameters used for algorithm testing

Algorithms	Parameters
RNN	Lookbacks 7, 14, 30, Units_list 16, 32, 64, Epochs_list 50, 60,70, 80, 90, 100, Batch_sizes 16,32,64 and Num_layers 1, 2, 3
LSTM	
GRU	

Table 2 shows the parameters used for algorithm testing.

2. Model performance evaluation was conducted by dividing the dataset into training and testing data. The training data, covering the period from 1 January 2020 to 31 December 2023, were used to develop models using three algorithms: RNN, LSTM, and GRU. The testing data, spanning from 1 January 2024 to 31 December 2024, were then used to evaluate the model's performance.

3.4 Evaluation

The forecasting performance of the models was evaluated and compared using five metrics, namely the lowest Mean Squared Error (MSE), the lowest Root Mean Squared Error (RMSE), the lowest Mean Absolute Error (MAE), the lowest Mean Absolute Deviation (MAD), and the lowest Mean Absolute Percentage Error (MAPE).

3.5 Forecasting

Based on the model evaluation results, the best-performing model was selected and subsequently employed to forecast hydrological inflows to the Xayaburi Dam in the Lao People's Democratic Republic for the next 365 days.

4 Experimental Results

The experimental results were divided into three parts, as follows:

4.1 Forecast Model Creation

Three techniques were used to develop forecasting models in the experiment: RNN, LSTM, and GRU. The evaluation results are presented in Tables 3, 4 and 5.

Table 3. RNN results

Parameters					Performance				
Look backs	Units list	Epochs list	Batch sizes	Num layers	MSE	RMSE	MAE	MAD	MAPE
14	64	100	16	1	0.4517	0.6721	0.2418	0.0719	4.9977
14	**64**	**100**	**32**	**1**	**0.4168**	**0.6456**	**0.2195**	**0.0637**	**4.3866**
14	64	100	64	1	0.5158	0.7182	0.2601	0.0724	5.2087
30	16	50	16	1	0.9753	0.9876	0.3257	0.0771	5.2937
30	16	60	32	1	1.4106	1.1877	0.3856	0.0735	5.9561
30	16	50	64	1	3.0273	1.7399	0.5910	0.1120	8.3324
30	16	60	16	1	0.8211	0.9061	0.3018	0.0755	5.2809

Table 3 presents the best evaluation results for the RNN model. The optimal parameter settings include a look-back value of 14, 64 units, 100 epochs, a batch size of 32, and one hidden layer. Under these settings, the RNN achieved an MSE of 0.4168, an RMSE of 0.6456, an MAE of 0.2195, an MAD of 0.0637, and an MAPE of 4.3866. The comparison between actual and predicted inflow values (m^3/s) for the RNN model is illustrated in Fig. 2.

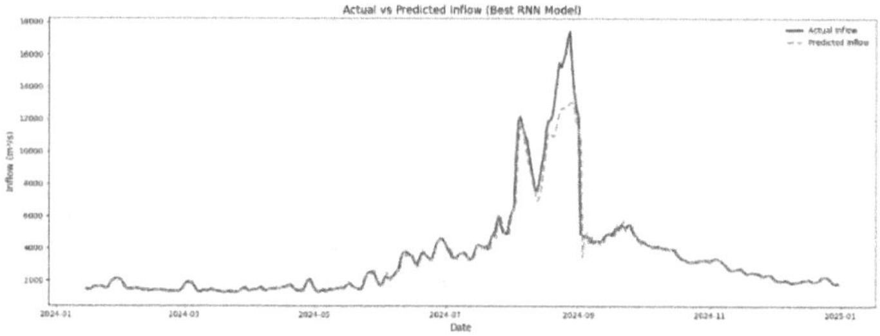

Fig. 2. Actual and predicted inflow values (m^3/s) for the RNN model.

Table 4. LSTM results

Parameters					Performance				
Look backs	Units list	Epochs list	Batch sizes	Num blayers	MSE	RMSE	MAE	MAD	MAPE
7	**64**	**90**	**32**	**1**	**0.3448**	**0.5872**	**0.1981**	**0.0608**	**4.3236**
7	64	90	64	1	0.4312	0.6567	0.2311	0.0863	5.1344
14	16	50	32	1	1.2524	1.1191	0.3650	0.0612	5.6764
14	16	60	64	1	2.0702	1.4388	0.4688	0.0705	6.8086
14	16	90	64	1	1.3826	1.1758	0.3864	0.0724	6.0957
14	16	100	16	1	0.6310	0.7944	0.2796	0.0850	5.6638
14	16	100	64	1	1.6188	1.2723	0.4283	0.0890	6.9026

Table 4 presents the best evaluation results for the LSTM model. The optimal parameter settings include a look-back value of 7, 64 units, 90 epochs, a batch size of 32, and one hidden layer. Under these settings, the LSTM achieved an MSE of 0.3448, an RMSE of 0.5872, an MAE of 0.1981, an MAD of 0.0608, and an MAPE of 4.3236. The comparison between actual and predicted inflow values (m^3/s) for the LSTM model is illustrated in Fig. 3.

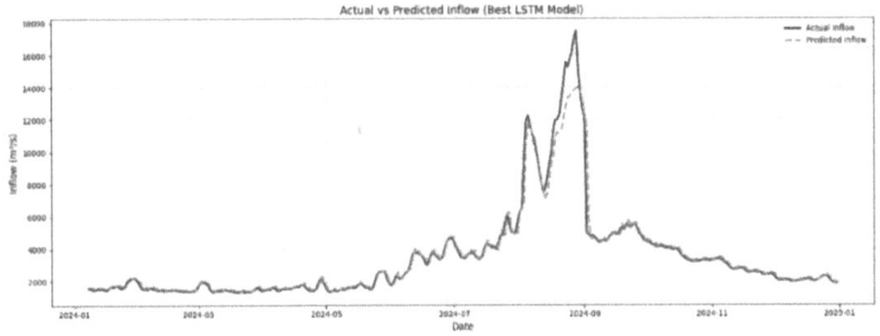

Fig. 3. Actual and predicted inflow values (m^3/s) for the LSTM model.

Table 5. GRU results

Parameters					Performance				
Look backs	Units list	Epochs list	Batch sizes	Num layers	MSE	RMSE	MAE	MAD	MAPE
7	32	70	64	1	0.9449	0.9721	0.3250	0.0851	5.6690
7	64	50	64	1	0.6176	0.7859	0.2834	0.0885	5.7504
7	64	70	16	1	0.3879	0.6228	0.2625	0.1309	6.4654
7	64	100	16	1	0.4060	0.6372	0.2447	0.0861	5.2644

(*continued*)

Table 5. *(continued)*

Parameters					Performance				
Look backs	Units list	Epochs list	Batch sizes	Num layers	MSE	RMSE	MAE	MAD	MAPE
7	**64**	**100**	**32**	**1**	**0.3757**	**0.6129**	**0.2252**	**0.0824**	**4.9454**
14	16	50	32	1	1.3840	1.1764	0.3831	0.0882	6.0397
14	16	80	64	1	2.4575	1.5676	0.5147	0.0905	7.4981

Table 5 presents the best evaluation results for the GRU model. The optimal parameter settings include a look-back value of 7, 64 units, 100 epochs, a batch size of 32, and one hidden layer. Under these settings, the GRU achieved an MSE of 0.3757, an RMSE of 0.6129, an MAE of 0.2252, an MAD of 0.0824, and an MAPE of 4.9454. The comparison between actual and predicted inflow values (m^3/s) for the GRU model is illustrated in Fig. 4.

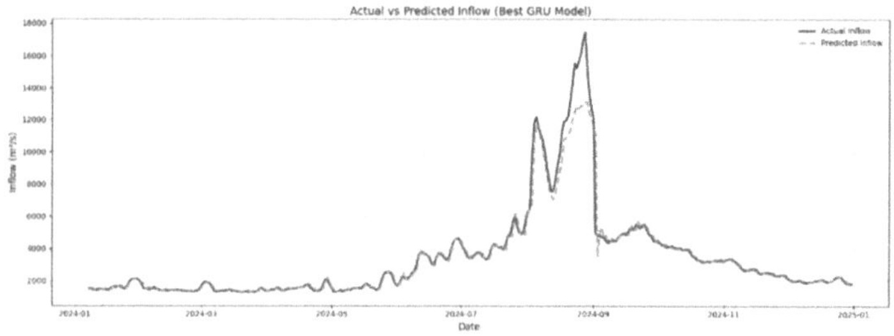

Fig. 4. Actual and predicted inflow values (m^3/s) for the GRU model.

4.2 Model Comparison and Selection

The forecast performances of the RNN, LSTM, and GRU models were compared using MSE, RMSE, MAE, MAD, and MAPE, as shown in Table 6.

Table 6. Model comparison results

Algorithms	MSE	RMSE	MAE	MAD	MAPE
RNN	0.4168	0.6456	0.2195	0.0637	4.3866
LSTM	**0.3448**	**0.5872**	**0.1981**	**0.0608**	**4.3236**
GRU	0.3757	0.6129	0.2252	0.0824	4.9454

Table 6 presents the comparison results among the three models. The LSTM model achieved the lowest MSE (0.3448), RMSE (0.5872), MAE (0.1981), MAD (0.0608), and MAPE (4.3236), indicating superior forecasting accuracy. The RNN model ranked second, while the GRU model performed comparatively worse. Therefore, the LSTM model was selected as the optimal model for subsequent inflow forecasting over the next 365 days.

4.3 Model Verification and Forecasting

After identifying the optimal model, hydrological inflows to the Xayaburi Dam in the Lao People's Democratic Republic were forecast for the next 365 days, from 1 January 2025 to 31 December 2025. The forecasting results generated using the LSTM algorithm is presented as an example in Table 7.

Table 7. Sample forecasting results of the LSTM model

Date	inflow (m^3/s)	Date	inflow (m^3/s)
01–01-25	2,365.49	22–12-25	2,580.45
02–01-25	2,365.49	23–12-25	2,645.77
03–01-25	2,519.56	24–12-25	2,722.21
04–01-25	2,596.33	25–12-25	2,786.38
05–01-25	2,597.02	26–12-25	2,810.94
06–01-25	2,535.29	27–12-25	2,745.76
07–01-25	2,437.44	28–12-25	2,619.34
08–01-25	2,332.52	29–12-25	2,486.13
⋮	⋮	⋮	⋮
21–12-25	2,529.74	31–12-25	2,409.70

Based on the results presented in Table 7. The forecasted inflow values for the period from 1 January 2025 to 31 December 2025 were combined with the historical inflow data from 1 January 2020 to 31 December 2024 to generate a consolidated time-series plot, as illustrated in Fig. 5.

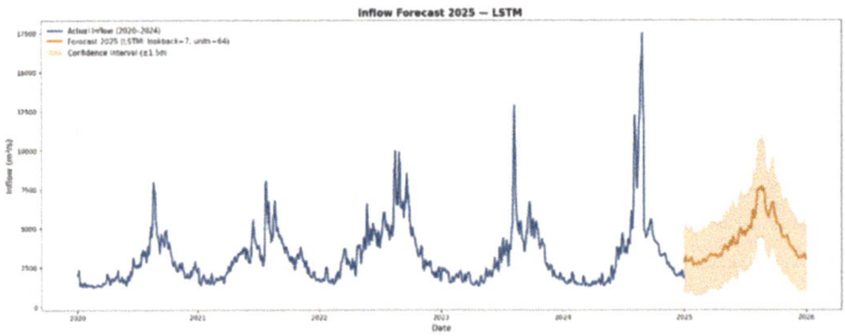

Fig. 5. Time-series plot of historical (2020–2024) and forecasted (2025) hydrological inflows to the Xayaburi Dam.

5 Conclusion

This study aimed to forecast hydrological inflows to the Xayaburi Dam in the Lao People's Democratic Republic using deep learning techniques. Daily data collected from 1 January 2020 to 31 December 2024, comprising 1,827 records, were utilized. The dataset was divided into training data (2020–2023) and testing data (2024). Three deep learning models, namely RNN, LSTM, and GRU, were developed and evaluated using five performance metrics: MSE, RMSE, MAE, MAD, and MAPE.

The experimental results indicate that the LSTM model achieved the best forecasting performance among the three models, with an MSE of 0.3448, RMSE of 0.5872, MAE of 0.1981, MAD of 0.0608, and MAPE of 4.3236. In comparison, the RNN model obtained an MSE of 0.4168, RMSE of 0.6456, MAE of 0.2195, MAD of 0.0637, and MAPE of 4.3866, while the GRU model yielded an MSE of 0.3757, RMSE of 0.6129, MAE of 0.2252, MAD of 0.0824, and MAPE of 4.9454. The results show that the LSTM model outperforms the RNN and GRU models because its architecture—specifically its memory cell and gating mechanisms (forget, input, and output gates)—effectively mitigates the vanishing gradient problem. This enables the model to capture complex, non-linear, and long-term temporal dependencies inherent in the Xayaburi Dam's hydrological data, leading to superior accuracy in multivariable time-series forecasting.

Consequently, the LSTM model was selected as the optimal model and subsequently applied to forecast hydrological inflows to the Xayaburi Dam over the next 365 days, from 1 January 2025 to 31 December 2025. The findings of this study confirm that LSTM-based forecasting effectively captures temporal dependencies in hydrological time-series data and supports dam operation and water resource management. Nevertheless, forecasting accuracy may be influenced by data availability and hydrological variability; therefore, future research may focus on incorporating additional climatic variables, longer historical records, or hybrid deep learning models to further improve forecasting performance.

References

1. Mahdi, N., et al.: Comparative evaluation of deep learning techniques in streamflow monthly prediction of the Zarrine River Basin. Water **16**, 208 (2024)
2. George, F., Alfred, K.: Implementing a GRU neural network for flood prediction in Ashland City, Tennessee. arXiv 2405, 10375 (2024)
3. Wani, O., et al.: Predicting rainfall using machine learning, deep learning, and time series models across an altitudinal gradient in the North-Western Himalayas. Sci. Rep. **14**, 27876 (2024)
4. Lipton, Z.C., Berkowitz, J., Elkan, C.: A critical review of recurrent neural networks for sequence learning. arXiv 1506.00019 (2015)
5. Krichen, M., Mihoub, A.: Long short-term memory networks: a comprehensive survey. AI **6**, 215 (2025)
6. Dey, D., Mridha, M.F., Saha, A., Babu, N.R.: A review of gated recurrent units in deep learning. IEEE Access **8**, 174006–174021 (2020)
7. Shi, J., et al.: Time series prediction model using LSTM transformer neural network for mine water inflow. Sci. Rep. **14**, 18284 (2024)

8. Ameri, R., et al.: Forecasting PM2.5 concentration based on integrating the CEEMDAN decomposition method with SVM and LSTM. Ecotoxicol. Environ. Safety **266**, 115572 (2023)
9. Sangkroa, J.: Development of a rainfall forecasting model for drought analysis in the Nan River Basin. Master's thesis, Department of Civil Engineering, Thammasat University (2023)
10. Klamsiri, T.: Development of a rainfall forecasting model for drought analysis in the Mun River Basin. Master's thesis, Department of Civil Engineering, Thammasat University (2023)

Author Index

S. Sodsee et al. (Eds.): IC2IT 2026, LNNS 5590, pp. 191–192, 2026.
https://doi.org/10.1007/978-3-032-28048-0

FSC
www.fsc.org
MIX
Papier | Fördert gute Waldnutzung
FSC® C083411

Kolibri 360 GmbH
Ferdinand-Jühlke-Straße 7
99095 Erfurt, Deutschland
produktsicherheit@kolibri360.de